CAMBRIDGE STUDIES IN ECONOMIC HISTORY
PUBLISHED WITH THE AID OF THE ELLEN McARTHUR FUND

GENERAL EDITOR
M. M. POSTAN
Professor of Economic History in the University of Cambridge

THE AGRICULTURAL REVOLUTION IN SOUTH LINCOLNSHIRE

THE AGRICULTURAL REVOLUTION IN SOUTH LINCOLNSHIRE

BY
DAVID GRIGG
Lecturer in Geography in the University of Sheffield

CAMBRIDGE
AT THE UNIVERSITY PRESS
1966

CAMBRIDGE UNIVERSITY PRESS
Cambridge, New York, Melbourne, Madrid, Cape Town, Singapore, São Paulo, Delhi

Cambridge University Press
The Edinburgh Building, Cambridge CB2 8RU, UK

Published in the United States of America by Cambridge University Press, New York

www.cambridge.org
Information on this title: www.cambridge.org/9780521106153

© Cambridge University Press 1966

This publication is in copyright. Subject to statutory exception
and to the provisions of relevant collective licensing agreements,
no reproduction of any part may take place without the written
permission of Cambridge University Press.

First published 1966
This digitally printed version 2009

A catalogue record for this publication is available from the British Library

Library of Congress Catalogue Card Number: 66–10798

ISBN 978-0-521-05152-1 hardback
ISBN 978-0-521-10615-3 paperback

FOR MY PARENTS

CONTENTS

List of maps and graphs	page ix
List of tables	xi
Acknowledgements	xiii
Abbreviations	xiv
Introduction	1

PART I. AGRICULTURE DURING THE NAPOLEONIC WARS

I	THE LAND	13
II	PRICES AND PROGRESS	33
III	FARMING METHODS AND PRODUCTIVITY	47
IV	THE PATTERN OF LAND USE	66
V	LANDLORDS, FARMERS AND FARMS	82
VI	THE AGRICULTURAL REGIONS	95

PART II. AGRICULTURE AFTER 1815

VII	PARADOX AND PROGRESS	117
VIII	THE AGE OF IMPROVEMENT	137
IX	THE NEW AGRICULTURAL GEOGRAPHY	155
X	THE CHANGING REGIONAL PATTERN	175
XI	SOME CONCLUSIONS	188

Bibliography	201
Index	207

LIST OF MAPS AND GRAPHS

1	South Lincolnshire	page 14
2	The simplified geology of South Lincolnshire	15
3	The parish boundaries of South Lincolnshire	16
4	The drainage of the fenland in the mid-nineteenth century	24
5	Indices of economic development, 1750–1820	34
6	Turnpike roads and navigable waterways	42
7	Areas of old enclosure and open field in 1790	51
8	Areas mainly under grass, 1795–1815	51
9	Yields of wheat, 1800 and 1851	61
10	The distribution of grain crops in 1801	73
11	The distribution of fodder crops in 1801	74
12	The percentage of the Land Tax paid by occupier owners, 1798–1808	84
13	The average size of holding, 1800	92
14	The agricultural regions in 1801	97
15	Wheat, beef and wool prices, 1811–80	118
16	Rental receipts on the Thorold, Welby and Ancaster estates	124
17	Tenant debt: the Thorold and Welby estates	124
18	Annual expenditure on the Thorold and Welby estates	130
19	Rent per acre in 1860	176
20	Percentage change in rent per acre, 1815 to 1840–7	177
21	England and Wales: rent per acre in 1815	195
22	England and Wales: percentage increase in rent per acre, 1815–60	196

LIST OF TABLES

1	Tolls on the Sleaford and Grantham waterways	page 35
2	Rentals in selected parishes, 1727–1814	36
3	Yield in bushels per acre, for wheat, barley and oats	59
4	The acreage in Kesteven under grain crops, 1792–1801	71
5	Number of quarters of grain exported from Boston, 1803–1815	71
6	The 1801 crop returns	72
7	Livestock in South Lincolnshire, 1798	79
8	Sheep densities in selected parishes, 1790–1815	80
9	Occupier owners, 1798–1812	86
10	The percentage of all holdings in each size group on the Ancaster estate, 1804 and 1830	91
11	Regional differences in farm-size structure	93
12	Farm size and landownership, 1813 and 1913	94
13	The agricultural regions, 1801	98
14	The average rent per acre in the major agricultural regions	99
15	Regional differentiation in rent per acre	102
16	Expenditure on C. Turnor's estate, 1830–93	131
17	Average crop yields, 1848–51	152
18	Quarters of wheat sold at Boston and Spalding markets, 1820–41	158
19	The regional pattern of land use, 1838–50	160
20	Coastwise shipments of grain from Boston, 1815–50	162
21	Regional livestock densities, 1875	165
22	Farm-size structure in Skirbeck Hundred, 1813 and 1870	168
23	Farm-size structure in Lincolnshire, 1851 and 1871	169

List of Tables

24	The regional pattern of farm size in 1851	page 170
25	The regional pattern of land use, 1800 to 1875	181
26	Percentage of total arable acreage in each region under grains and fodder crops, 1801 and 1875	182
27	Regional livestock densities, 1798 and 1875	184
28	Wheat yields in South Lincolnshire, 1870–8	186
29	Rent per acre in England and Wales, 1815–1912	197

ACKNOWLEDGEMENTS

I should like to acknowledge the help I have received during the preparation of this book from a great many people. First, the encouragement and advice I received from the two supervisors of the Ph.D. thesis on which this book is based, Mr C. T. Smith and Dr H. C. K. Henderson. Dr D. R. Mills helped me greatly in the early stages and I would like to thank too Mr W. H. Hosford, who placed a number of documents at my disposal. Professor K. C. Edwards, Professor M. M. Postan and Dr Jean Mitchell all made most helpful criticisms of an early draft of the text.

Much of the text is based on documents in a number of record offices, and I benefited greatly from the assistance given by officials at the Public Record Office, the Ministry of Agriculture at Guildford, Lincoln Public Library, Kesteven County Library and Holland County Council. Mr S. W. Woodward, Secretary of Spalding Gentlemen's Society, gave me ready access to the records of his society. But my greatest debt in this way is to Mrs Joan Varley and her colleagues at Lincoln Archives Office who were most generous in their assistance.

Lastly I should like to thank Mr J. Hall and Miss S. Ottewell who drew most of the maps, and Miss Zena Oxley who typed the manuscript.

A number of the maps in this book have already appeared in print and I am grateful to the Editors of *The East Midland Geographer*, *The Transactions of the Institute of British Geographers*, *Agricultural History* and *The Agricultural History Review* for permission to reproduce them here.

An earlier draft of this book was awarded an Ellen McArthur Prize in 1961 and the book is published with the aid of the Ellen McArthur Fund.

D.B.G.

Stephenson Hall, University of Sheffield
September 1963

ABBREVIATIONS

The following abbreviations have been used in the footnotes:

L.A.O.	Lincoln Archives Office
S.G.S.	Spalding Gentlemen's Society
P.R.O.	Public Record Office
Annals	*Annals of Agriculture*
B.P.P.	British Parliamentary Papers
J.R.A.S.E.	*Journal of the Royal Agricultural Society*

INTRODUCTION

A comparison between English farming in 1750 and 1850 reveals a number of striking differences. Yet they were differences of degree, rather than kind, for the agricultural revolution was brought about by the acceleration of existing trends rather than any sudden innovations. Indeed, an early Victorian farm would have presented few surprises—with the exception of the reaper and the steam threshing machine—to a good mid-eighteenth century farmer.

Perhaps the most far-reaching changes came in farm organization, and certainly no other topic received so much attention from contemporary writers. In 1750, in spite of several centuries of enclosing, about half the cultivated area of England was still farmed under the open-field system.[1] Farms consisted of a number of strips, scattered among the open fields, which were unfenced and subject not only to a common rotation but to common grazing after each harvest. The enclosure of the open fields, which was nearly complete by the end of the Napoleonic Wars,[2] allowed the consolidation of a farmer's holdings into one compact unit. The reorganization brought about by enclosure sometimes led to the amalgamation of small holdings into larger farms; but 'engrossing', as this process was called, was probably less common than eighteenth-century pamphleteers supposed. Changes in landownership during enclosure are less easy to trace. In some cases the smaller landowners—particularly the occupier owners—finding their share of the cost of enclosure excessive, may have sold out to larger landowners. But the decline of the occupier owner, which was once attributed primarily to enclosure seems to have been largely completed before the era of Parliamentary enclosure. Certainly by the 1780's the occupier owner was the exception and the tenant farmer the rule in English farming.[3]

[1] W. H. R. Curtler, *The Enclosure and Redistribution of our Land* (Oxford, 1920), p. 149.
[2] E. K. C. Gonner, *Common Land and Inclosure* (London, 1912), p. 2. In 1820 only two counties, Oxford and Cambridge, had more than 5 per cent of their total area remaining in common field.
[3] E. Davies, 'The small landowner, 1780–1832, in the light of the Land Tax assessments', *Economic History Review*, I, no. i (1927), 111–12.

The Agricultural Revolution in South Lincolnshire

Although the introduction of new crops and methods was sometimes possible in the open fields,[1] their survival was generally a deterrent to the adoption of new techniques. It was the technical progress of the late eighteenth century and the first half of the nineteenth century which led to the increases in crop and livestock productivity, which are perhaps the most important feature of the agricultural revolution. Between 1750 and 1850 the average wheat yield in England nearly doubled, a rate of increase greater than in any preceding period.[2] This was achieved, not simply by the introduction of fertilizers and new rotations but by a fundamental change in the farming system. In the open fields, animal and crop production were carried on together on the same farm but were only partially integrated. During the eighteenth century, and increasingly in the nineteenth century, the adoption of certain new farming methods not only increased crop productivity, but firmly integrated livestock and crop husbandry. It was this 'mixed farming' which was the culmination of a series of advances in agriculture which had begun in the seventeenth century, and which were the basis of the 'High' farming of eastern England in Victorian times. This latter period can be said to mark the end of the agricultural revolution, not because there ceased to be the advances in technique, but because the changes were different in *kind* from those of the earlier nineteenth century.

The fundamental improvement of the eighteenth century was the replacement of the normal open-field rotation of two grain crops and a bare fallow. Of the new rotations the Norfolk four course was justly the most celebrated.[3] The turnip crop not only utilized the fallow and provided additional fodder, but when it was properly cultivated with the drill and hoe proved an excellent cleaning crop. In some areas, noticeably on light soils, sheep were folded on the crop and their feet consolidated the soil whilst their manure enriched the land for the following wheat crop. The growth of temporary grasses provided grazing and hay, and, perhaps more

[1] M. Havinden, 'Agricultural progress in open field Oxfordshire', *Agricultural History Review*, IX, part ii (Oxford, 1961), 73–83.
[2] M. K. Bennet, 'British wheat yields per acre for seven centuries', *Economic History*, III, no. 10 (1935), 26, 28.
[3] The development of Norfolk farming is described by N. Riches in *The Agricultural Revolution in Norfolk* (Chapel Hill, North Carolina, 1937).

Introduction

important, increased the nitrogen content of the soil. The new rotations alone would have increased crop yields and the stock-carrying capacity of the land. But there were other improvements. Cattle were increasingly stall fed on both crops and purchased oil-cake; this both fattened the animals more efficiently and gave a greatly enriched manure. In addition to farmyard manure, artificial fertilizers such as bones, guano and lime all contributed to increased crop yields. On heavy soils some of these innovations were impossible without adequate drainage, and thus in the clay lands of England underdrainage was essential. The progress in arable farming was paralleled in livestock husbandry by increases in both the quantity and quality of milk, wool and meat.

But the rise in the total output of English farming was not accomplished solely by greater outputs per acre and per animal. The enclosure of heath and moorland together with the reclamation of fen and marsh added greatly to the cultivated area. Although there are no reliable statistics on the total agricultural area of England until the 1860's, one authority has estimated that there was an increase of nearly a third between 1700 and 1854.[1] This came from land which had long lain waste, but there was also an increase in the arable acreage at the expense of grassland, for the spread of mixed farming meant that livestock were fed increasingly on fodder crops and artificial feeds. By the 1860's permanent grassland remained in eastern England only where the land was either unsuited to arable or gave a particularly fine pasture.

Whilst the agricultural revolution saw considerable advances in crop and livestock productivity, there was less progress in labour productivity. Indeed rather than causing a fall in labour needs the New Husbandry almost certainly required an increase. The process of enclosure, with its construction of fences, new roads and farmhouses, required a considerable if temporary work-force. So too did the reclamation of fen, marsh and heath which was going on in many parts of England and Wales.[2] Arable farming needed more labour

[1] L. Drescher, 'The development of agricultural production in Great Britain and Ireland from the early nineteenth century', *The Manchester School of Economic and Social Studies*, XXIII, no. 2 (1955), 167.
[2] For example the reclamation of Traeth Mawr employed over three hundred men for six years from 1805 to 1811 (A. H. Dodd, *The Industrial Revolution in North Wales*, London, 1933, p. 43).

per acre than grazing, and the first half of the nineteenth century saw an expansion of arable not only into former waste land but into areas of permanent pasture. Under the Norfolk system the land was worked more frequently and more carefully than on the open fields. The turnip crop alone required an enormous amount of labour if it was to be properly drilled and hoed. The increasing rural population of this period meant, however, that there was no shortage of labour, except perhaps locally, and labour was used prodigally in places. Thus in several limestone areas gangs of women and children were used for stone-picking.[1] The Norfolk system, and its regional variants, were based on cheap labour, and the use of machinery was not such an important part of the agricultural revolution as it was in the industrial revolution. It is true that new implements were being used, the drill being perhaps the most important whilst the plough had been greatly improved in some areas. But the drill was not primarily a labour-saving device, and such implements were few in the first half of the nineteenth century. Although the steam threshing machine, the reaper and other devices were all being used in the 1820's and 1830's their general adoption came much later.[2]

The great changes in agriculture did not of course occur independently of the rest of the economy, and were but part of the accelerated growth in the British economy which began in the later eighteenth century. Much of the stimulus to agricultural development came from the increased demand for agricultural products from the growing urban population. This, coupled with scarcity during the Napoleonic Wars, made agricultural prices more favourable in the second half of the eighteenth century than in the first. Many of the changes in agriculture, and indeed in industry, would have been impossible without the improvements made in transport by the construction of canals and the turn-piking of roads. Not only were remote areas linked to new markets for the first time, but existing facilities were improved so that the movement of bulk commodities was cheapened. This allowed farmers to use lime, bones, tiles for underdrainage and later coal for steam engines. The market-

---

[1] W. Hasbach, *A History of the English Agricultural Labourer* (London, 1908), p. 195.
[2] Sir John Clapham, *An Economic History of Modern Britain* (Cambridge, 1926), I, 140–2.

Introduction

ing of farm products was also made easier and this not only encouraged the expansion of output, but by freeing the farmer from simply local demand, led to changes in the type of output.

The improvements in agriculture required considerable expenditure. Unfortunately there is little reliable evidence on the sources of capital. Landowners, rather than tenants, were responsible for many of the costlier improvements—enclosure, fen drainage and marsh reclamation. It does not seem that in the later eighteenth century there was such a flow of capital into English farming from non-agricultural sources as there had been at an earlier date.[1] To some extent the capital was self-generated, for both tenants and landlords must have made considerable profits during the inflationary period at the end of the century. Much of this profit was probably reinvested in the land. One factor which must have encouraged investment—by tenants particularly—was the rapid growth of country banking in the last two or three decades of the century.[2]

English farming contributed to the general economic growth as well as benefiting from developments in other industries. Increased agricultural output provided food for the growing urban population although it is true that in the nineteenth century the country became increasingly reliant on imports. Further, the new industrial towns must have acquired much of their labour force from rural migration, although the complex relationship between enclosure, increasing farm productivity and rural migration is not altogether clear.[3] Agriculture directly contributed to industrial growth by providing markets for a number of new industries; the manufacture of farm implements is an obvious example.

This account of the agricultural revolution in England has necessarily been brief and generalized, and many of the topics mentioned are still the subject of considerable controversy. Thus the relationship between enclosure, the occupier owner and increasing farm size is far from settled. When exactly did the occupier owner dis-

---

[1] H. Habbakuk, 'English landownership, 1680–1740', *Economic History Review*, XI, no. i (1940), 4–5.
[2] L. S. Pressnell, *Country Banking in the Industrial Revolution* (Oxford, 1956).
[3] For a recent analysis see J. D. Chambers, 'Enclosure and Labour Supply in the Industrial Revolution', *Economic History Review*, v, no. iii (1953), 319–43.

appear from English agriculture? Was enclosure always followed by the amalgamation of small holdings? If not, did engrossing occur in the eighteenth century in areas which had been enclosed in previous centuries? A series of unresolved problems surrounds the question of technical progress. Recent work has shown that many of the new methods associated by early writers with the period of Parliamentary enclosure were already being practised in some parts of England at a much earlier date. If this is so, what factors controlled the diffusion of new techniques? Has the survival of the open fields as an obstacle to technical progress been overestimated? Is there any correlation between farm size and landownership and the rate of technical progress? Other aspects of the agricultural revolution have as yet received little attention. How, for instance, did the new methods affect land use? And conversely to what extent did existing patterns of land use affect the adoption of new techniques?

To some extent it is not surprising that these and other topics in a much discussed period remain unresolved, for the evidence is incomplete and often unreliable. In the first place there is a dearth of reliable statistical information on many of the points at issue. For example many of the problems of technical progress would be solved if there was comprehensive information on crop yields. On crop and livestock numbers there are only estimates for the whole country until the Board of Agriculture's returns begin in 1866, with the exception of the incomplete parochial returns made to the Home Office in 1801. On farm sizes there are only scattered estate surveys for small areas until the 1851 census; and even these latter accounts are far from exact. Landownership can only be approached on a regional scale, by using estate records and the Land Tax returns. The latter whilst forming the basis of much work on the occupier owner, can be no more than a rough guide. The neglected topic of land use changes presents particular problems, for either maps or statistics on a parish basis are essential. Before the Board of Agriculture's returns, only Tithe Maps and the 1801 returns give any quantitative indication of the distribution of arable, grass and rough grazing.

Literary evidence, although more abundant than statistical

Introduction

evidence, is not without its defects, and the record is far from complete. Archive collections may have estate and farm papers which describe farming policy and methods, but these are invariably sporadic in their survival both in time and place, and form a dubious basis for generalization. This type of evidence, for instance, makes it relatively easy to date the first appearance of a new technique or crop in an area; but it does not tell us when it was *generally* adopted. It is this difficulty which has led to much controversy on the rate of technical progress in the eighteenth and nineteenth centuries. It is possible to supplement the records of archive offices with the descriptions of agriculture to be found in contemporary topographies and tours, whilst at the end of the eighteenth century there is a rich descriptive literature, culminating in the *General Views of the Agriculture* of each county. But the writers of these detailed reports were often strangers to the county they described, and many of them were concerned only with describing the best practices rather than average farming conditions.

A second difficulty in interpreting the changes of the agricultural revolution lies in the pronounced regional differences in farming practice which have always existed in England. What can be shown to be true for one part is not necessarily true for another. To be sure the farming of the whole country was uniformly subject to certain external economic and political factors. But the differences between farming regions—in landownership and farm size as well as in soil and land use—meant that they may respond in a very different manner.

Thus most of the new farming methods adopted in the eighteenth century were in arable farming. Yet in the late eighteenth century arable land predominated only in the east and south of England. The new techniques then had relatively little to contribute to the farming of much of the Midlands and the west. Even within the arable areas of the east different regions responded in very different ways. The Norfolk system of farming developed on the light sand soils of that county, and was not easily adopted on clay soils. Thus throughout much of the later eighteenth and early nineteenth centuries there were striking contrasts between the development of agriculture on the limestone uplands of Yorkshire and Lincolnshire and the low, badly drained clay vales of the same counties. Farming regions, of course, differed in other ways. Enclosure was held by eighteenth-century writers to have been the spearhead of progress, and without the stimulus of enclosure little was done. Yet in the early eighteenth century much of England was already enclosed. It would be profitable to compare the rate of agricultural development in old enclosed districts with that in late enclosed but otherwise similar regions to assess the significance of enclosure properly.

This book has been written in order to suggest an approach to some of these problems, particularly that of regional change. It deals with the changes which took place in the farming of South Lincolnshire[1] during the agricultural revolution of the late eighteenth and early nineteenth centuries. But simply to confine the study to a county—or a part of a county—is not an answer to the problem of regional change. This study is concerned more with the regional differences *within* South Lincolnshire than with the contrasts between the area and the rest of England, though these are necessarily touched upon.

A regional discussion of this type requires a geographical approach. Consequently much of the book is concerned with first, how the agricultural geography of the area varied regionally, and how it changed over a period of time. Ideally a study of regional change during the agricultural revolution should begin in the early eighteenth century. But unfortunately there is not sufficient literary or statistical evidence adequately to describe the regional differences within such a relatively small area, or to be able to distinguish the average conditions from the best farming practices. Indeed it is not until the end of the century that this is possible, and so the first part of the book deals with the agriculture of South Lincolnshire during the Napoleonic Wars. The agricultural geography of the area is described both systematically and regionally; and the factors influencing regional variations are separately analysed. Because this was a period of apparently rapid change, attention is also paid to the rate of change—for instance the rate at which farm improvements were adopted and the expansion of the arable acreage. The second part of the book deals with regional changes which occurred after

[1] South Lincolnshire consists of the two Parts of Kesteven and Holland.

Introduction

the Napoleonic Wars, and the technical and economic conditions which influenced these changes. These developments are traced down to the middle of the century. By the 1860's 'High' farming was well established in most parts of South Lincolnshire, and the agricultural revolution can be said to have been completed. This is not to say that progress ended. Rather that new types of changes—particularly the introduction of labour-saving machinery and later the adaptations to falling prices—replaced the old.

The study begins with an analysis of the factors which determined both the regional differences in farming and the rate at which improvement was progressing in the late eighteenth century. A variety of factors is operative: the survival of open fields, differences in land tenure and farm size, accessibility—and so forth. But paramount was the condition of the land the farmers had to work, and so the first chapter deals with the soils of South Lincolnshire and the state of drainage in the fenland.

PART I
AGRICULTURE DURING THE NAPOLEONIC WARS

CHAPTER I

THE LAND

'Ah, Lincolnshire', George III is reputed to have said, 'all fens, flats and fogs!' Certainly the county has never been celebrated for its scenic beauty, and this may be because of the lack of any marked contrasts in relief. Nevertheless, the difference between the two upland areas, the Heath and the Wolds, and the extensive lowlands of the Vale of Trent, the Ancholme Valley and the fenland has exercised an enduring influence on drainage, settlement and farming. These broad contrasts in relief reflect fairly faithfully differences in geology whilst the geological make-up of the county has in turn profoundly influenced soil types. Even today the farming regions of Lincolnshire correspond closely to differences in soil.[1] In the eighteenth century, when farming methods were still primitive the dependence of the farmer on the soil was even greater.

Most of South Lincolnshire lies below 200 feet above sea-level, and only in the south-west does the land rise above 400 feet (Fig. 1). The principal land-form of the area is the cuesta of the Lincolnshire Limestone. This forms a west-facing scarp which runs south from Lincoln to the Ancaster Gap, and then south-westwards to the Leicestershire border, giving rise to an impressive cliff-like feature throughout its course (Figs. 1 and 2). The dip-slope of the cuesta is made up of the Lincolnshire Limestone together with a succeeding Limestone, the Cornbrash. North of the Ancaster Gap this limestone surface slopes gently eastwards and is scored by a number of dry valleys. This region, together with the dip-slope north of Lincoln, has long been known as Lincoln Heath.

South of the Ancaster Gap a number of geological changes alter the scenery.[2] Near Lincoln the Marlstone Ironstone forms no more

[1] See, for example, L. D. Stamp, *The Land of Britain*, parts 76 and 77 (London, 1942). For a more recent discussion, see K. C. Edwards, 'Lincolnshire' in J. Mitchell (ed.), *Great Britain: Geographical Essays* (Cambridge, 1962), pp. 308–29.
[2] H. Swinnerton and P. E. Kent, *The Geology of Lincolnshire* (Lincoln, 1949); D. L. Linton, 'The landforms of Lincolnshire', *Geography*, XXXIX, no. 2 (1954), 67–78.

13

The Agricultural Revolution in South Lincolnshire

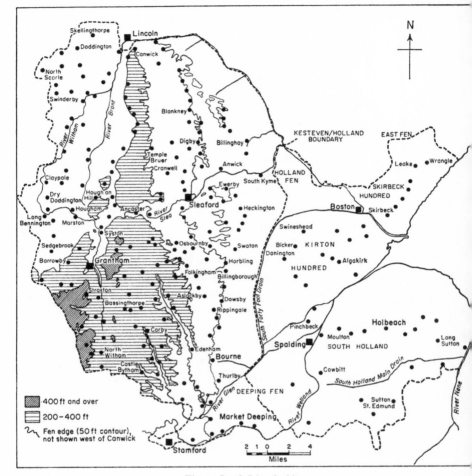

Fig. 1. South Lincolnshire.

than a narrow ledge at the base of the Limestone scarp, but southwards its lithology changes and it becomes a much more impressive feature. From Hough on the Hill to the Leicestershire border the Marlstone forms a second scarp running closely parallel to the Limestone scarp (Fig. 2). In addition the angle of dip of the Limestone lessens southwards so that whereas north of the Ancaster Gap the Heath has the classic form of a cuesta, to the south it is more

The Land

plateau-like. Finally, and of most importance, whilst the Heath north of the Ancaster Gap is largely free from glacial deposits,[1] to the south a considerable area is covered by Chalky Boulder clay. This completely alters the landscape and also presents very different farming problems. The limestone Heath proper only extends south

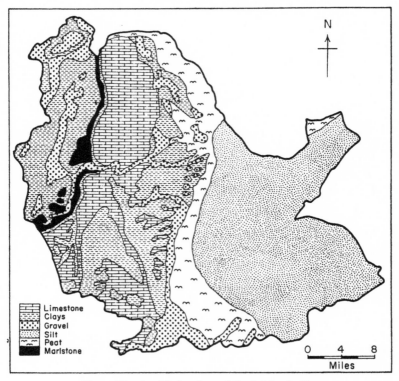

Fig. 2. The simplified geology of South Lincolnshire.

of the Gap in a narrow band in the west, although in the south between Stamford and South Witham the limestone is also free from boulder clay.

West of the Heath and its two scarps lies the low plain of west Kesteven, formed from Lias clays. The region is crossed by both the Witham, on its devious course to the Wash, and the Brant, which

[1] No drift deposits are marked on the geological map but a number of small deposits have since been identified.

joins the Witham above Lincoln. The low gradient of these two rivers has meant that in the past they have often flooded. In the north of the plain much of the Lias clay is covered by gravels deposited by the Proto-Trent.[1] To the west and south-west of

Fig. 3. The parish boundaries of South Lincolnshire (parts of Kesteven and Holland). B, Boston; BE, Bourne; SP, Spalding; L, Lincoln; G, Grantham; S, Sleaford; ST, Stamford.

Lincoln these deposits form a number of low ridges sometimes called the Graffoe Hills.[2]

The Heath dips gently eastwards towards the fenland; the postglacial deposits of silt and peat have largely covered the Oxford

[1] K. M. Clayton, 'The differentiation of the glacial drifts of the East Midlands', *The East Midland Geographer*, no. 7 (1957), 37–8.
[2] D. R. Mills, 'Regions of Kesteven devised for the purpose of agricultural history', *Reports and Papers of the Lincolnshire Architectural and Archaeological Society*, VII, part 1 (1957), 60–82.

The Land

clays which succeed the Jurassic limestones, so that the broad clay outcrop of the Ancholme Valley in Lindsey has no equivalent in South Lincolnshire. The clays form only a narrow zone between the limestones and the fen deposits, and are further concealed by a discontinuous stretch of gravels, which run the whole length of the fen edge from Lincoln to Market Deeping.

The fenland forms a sharp contrast to the rest of the area. Most of it lies below the level of high tide in the Wash and for the greater part it is a monotonous and undifferentiated plain. The major geological contrast within the fenland is between the peats and the silts. The silts, which vary considerably in texture, were deposited by tides coming into the proto-Wash, eventually forming a discontinuous bar. Behind this bar the rivers of the upland flowed into a lagoon and it was under these conditions that the peats were formed.[1] The peat deposits are much more extensive to the south in Cambridgeshire and Norfolk, and in South Lincolnshire they only occur in a narrow zone on the landward edge of the fen. Peat deposits also occur in the East, West and Wildmore Fens, which now lie mainly in Lindsey, but in the eighteenth century were intercommoned by a number of Holland villages.

Most of the Lincolnshire peatland is now to be found in Kesteven, from Lincoln in the north to Market Deeping in the south. It extends only a little over the boundary into Holland, although before 1850 it may have covered a more extensive area. Nevertheless, most of the Lincolnshire fenland now consists of silts. These are generally lightest near the coast, where deposition is most recent, and heaviest inland. The most significant feature of the silts is the presence of a broad low ridge running approximately parallel with the shore. Although no more than 3–5 feet above the silts and peat further inland it has had considerable consequences for the settlement, drainage and land-use history of the fenland.

Relief (by determining drainage) has only had an indirect influence on farming in the area. Of greater importance are the regional variations in soil type. Unfortunately, there is as yet no map of the

---

[1] S. J. Skertchley, *The Geology of the Fenland* (London, 1877); H. Godwin and M. Clifford, 'Studies in the post-glacial history of British Vegetation', *Transactions of the Royal Society: Philosophical Transactions*, B, CCXXX (1938).

soils of South Lincolnshire. However, as the parent materials are the major factor causing differences in soils within the area, the geological map (Fig. 2, p. 15) gives an approximate guide to soil types. Three major categories can be recognized. First, the 'light' soils developed on the gravels of north-west Kesteven, the limestones and the Marlstone; secondly the 'heavy' clay and clay loam soils developed on the Lias clay of west Kesteven, the Chalky Boulder clay of south-east Kesteven, and the small outcrops of Oxford clay in east Kesteven; and thirdly the gravel, peat and silt soils of the fenland.

Clay soils remain amongst the least rewarding of English agricultural soils. Their impermeability makes them difficult to drain, and they are costly to work because of their heavy texture. When imperfectly drained they easily become water-logged. This keeps the root systems shallow and deprives the crop of plant nutrients. In winter the soils 'heave' during frost, breaking the roots. In spring much of the sun's energy is used up in evaporating the excess moisture rather than warming the soil, so that they have a shorter growing season than light soils. Not only are the clays difficult to cultivate, but a good tilth is not easily maintained. If the clays are cultivated when the soil moisture exceeds certain limits, large plastic lumps form; conversely in hot dry spells, the surface hardens and cracks. The fact that clays cannot be worked in wet weather shortens the working year and often delays sowing and harvesting with adverse effects on crop yields.[1]

These disadvantages have been reduced in modern times by better underdrainage and the introduction of caterpillar tractors, but prior to these improvements the clays were problem soils. A bare fallow was regarded as essential until quite recently, and the high cost of working them caused many farmers to keep their claylands under permanent grass. In the eighteenth century they presented particular problems. The Norfolk four course, and the folding of sheep, was a system devised for light lands and could not easily be adopted on undrained clays.

Turnips yielded poorly on clays, and even if they grew well, sheep could not be successfully folded on the crop; their feet 'poached'

[1] E. J. Russell, *Soil Condition and Plant Growth* (London, 1956), pp. 596–9.

The Land

the soil and they became liable to foot rot. The turnips could of course be lifted and fed off elsewhere, but this was often prohibitively expensive in wet weather, and in any case the land lost the benefit of manuring. Whilst the clays remained undrained manures were easily washed away, and so there was little incentive to use artificial fertilizers.

Even when drained the clays were physically difficult and thus costly to work because of their heavy texture. However, the clays are not without some advantages. There is little difficulty in maintaining the humus content, and clay soils give good yields of wheat oats, beans and grass. Roots, as has been noticed, are not successful, nor is malting barley, because of the short growing season.[1] The clay soils of South Lincolnshire have all the characteristics described above, but differ slightly from each other.[2] The stiff, blue Lias clays give a heavy, brown clay or clay loam soil, deep and fertile, but difficult to till. Much of this land has long been under grass, although the pastures are acid and need liming. Acidity is less of a problem on the Chalky Boulder clays, for the lime content is maintained by the limestone fragments in the clay. The soils on the Chalky Boulder clay are less homogeneous than those on the Lias, but are all heavy in texture and badly drained. Generally bluish grey in colour, they have a varying admixture of stone and sand. The small outcrop of Oxford clay to the east and north-east of the boulder clays gives a grey brown soil with a thick, stiff, yellow subsoil which impedes drainage.

The light soils—those developed on the gravels of north-west Kesteven, the thin outcrop of the Marlstone, and the limestones of the Heath—differ in nearly every respect from the heavy soils. The soils are free draining, and the problem is not to remove excess moisture but to maintain enough in the seed bed.[3] In summer the free drainage of the limestone combined with evaporation 'scorches' grasses, and the light soils consequently rarely give a good feeding pasture. Their light texture means they are easily worked but it also

---

[1] G. V. Jacks, *Soil* (London, 1954), pp. 21–3.
[2] K. C. Edwards, 'The soils of the East Midlands', in C. E. Marshall, *Guide to the Geology of the East Midlands* (Nottingham, 1948), pp. 86–9.
[3] Russell, pp. 595–6.

encourages weeds. The limestone and Marlstone soils are rich in mineral foods but deficient in organic matter. In the eighteenth and nineteenth centuries the growth of green crops and the folding of sheep helped to maintain the humus content of the soil. In addition the sheep's feet consolidated the loose surface soil. The light soils of South Lincolnshire are not identical. The gravels give a shallow, light, stony soil made up of sand and silt. The limestones give a thin, reddish, sandy loam with a high proportion of angular fragments; they are generally stonier near the crest of the cuesta than farther down the dip-slope and are also stonier south of the Ancaster Gap. The Marlstone, unlike the limestone or the gravels, gives a deep soil, which is reddish in colour, friable and easily worked.[1]

The fenland soils although complex and in some cases even changing radically within one field, can be divided into three categories: the gravel soils, the peat soils and the silts.

Gravels occur on the margin of the Lincolnshire fen almost continuously from Lincoln to Market Deeping. There is some dispute as to their origin, but it seems likely that those from Lincoln to Heckington were deposited by an earlier course of the River Witham and those from Heckington south are partly of marine and partly of riverine origin. Only near Market Deeping do the gravels cover a significant area, and they give rise to a deep, sandy loam which is easily cultivated and drained.[2]

The peat deposits lie landward of the silts. In Lincolnshire the peat deposits are thin and interdigit with the silt deposits to the east, so that there is no distinct boundary. Not only are the peats thin, but their depth varies from place to place. Everywhere, however, they are underlain by clay. Before the peats were drained and cultivated they were certainly thicker, and may also have covered a greater area. Peat in its natural state consists largely of water, thus as the fen was progressively better drained, the surface of the peats shrank; further the peat has in many places been mixed with the

[1] Edwards, *loc. cit.* pp. 87, 90–1; *Journal of the Royal Agricultural Society*, XII (1851), 259 ff.

[2] J. E. Prentice, 'The subsurface geology of the Lincolnshire fenland', *Transactions of the Lincolnshire Naturalists Union* (Louth, 1950), 12, 136; G. Smith, *The Land of Britain* (London, 1937), part 69, 11–12.

The Land

underlying clays, so losing its initial characteristics.[1] At present the peats are confined to the Kesteven fenland, with small extensions into Holland in Deeping Fen and East Fen.

Peat soils are black or dark brown and are highly friable, with no well-developed soil profile. As long as they have an adequate surface drainage they drain easily; indeed in a dry summer they often have an inadequate moisture content and become susceptible to 'blowing'. Unlike the upland peats they are neutral in reaction and have a very high organic content, which accounts for their fertility. When the peats were first drained and cultivated the excessive nitrogen content resulted in a very poor quality wheat, more straw than grain. A wide range of crops can be grown on the peats, but turnips have never been successful because of the tendency of the roots to go 'fangy'.[2]

Silt soils make up 90 per cent of the area of Holland. They vary from a fine sand to heavy clay. Three soil groups have been recognized on the basis of texture: the light silts, the medium and heavy silts, and the clay and skirty soils. The light silts are easily cultivated and are highly porous, but have a low organic content. They occur particularly near the coast. These soils need more careful cultivation than the other fen soils. The medium and heavy silts are predominantly dark brown in colour, less freely draining than the light silts, and with a higher organic content. The clay or skirty soils occur particularly in South Holland. They have a high clay and silt content, abundant organic matter, but are extremely stiff and difficult to drain. They have to be farmed in much the same way as the clay soils of Kesteven.[3]

In spite of the complex soil geography of the fenland, three regions have long been recognized both by the inhabitants and by agricultural writers. Ultimately these divisions are a function of differences in relief, but they have profoundly influenced the drainage, settlement, enclosure and land use of the area. The low ridge of silt running parallel with the shore has already been noticed. Standing

[1] G. Fowler, 'Shrinkage of the peat-covered fenlands', *Geographical Journal*, LXXXI, no. 2 (1933), pp. 149–50.
[2] G. Smith, pp. 12–13; A. Woodruffe Peacock, *Fenland Soils, Lincolnshire Naturalists Union* (Spalding, 1901).
[3] G. Smith, pp. 13–16.

The Agricultural Revolution in South Lincolnshire

5 or 6 feet above the landward fen, this ridge was the site of the earliest settlement in the fenland because it was relatively free from flooding.[1] The land around the villages was enclosed at an early date, and drains and roads follow winding courses. Most of the Townland, as the ridge is called, consists of light, medium and heavy silts. Landward of the Townland lies the Interior Fen. This consists of the whole of the peats and some of the lower and heavier silts. This area was subject to regular flooding until the nineteenth century and was largely held in common by the Townland villages until their enclosure in the late eighteenth century. Roads and drains nearly all date from this period and are straight and so the difference between the Townland and the Interior Fens is very clearly seen on an Ordnance Survey map. Seaward of the Townland, silt has accreted over a number of centuries and this new land has been embanked and enclosed at various times. The Marsh, like the Interior Fens, lacks nucleated settlement, and was generally enclosed at a later date than the Townland. Made up of a variety of silt soils, it does, however, have more light silts than the other two regions. The parish boundaries of the fenland, centred on the Townland settlements, are so drawn as to include a portion of fen, Townland and Marsh; they run at approximately right angles to the shore (Fig. 3, p. 16).

The main concern of the Heath farmer was to maintain the humus content of his soils. On the clays of Kesteven successful farming was impossible without efficient underdrainage. But throughout the fenland farming was dominated by the problem of surface drainage. Until the nineteenth century the rivers and drains of the fens flooded at regular intervals, whilst more irregularly but more disastrously much of the fenland was swamped by tidal waters which had breached the coastal banks. Thus it can be seen that there were very marked regional contrasts in the environment of South Lincolnshire; these differences have had a profound influence on the development of agriculture. By far the most intractable problem was the drainage of the fenland, and it is this which will be considered in the rest of this chapter.

[1] H. E. Hallam, 'The New Lands of Elloe', *Leicester University, Department of Local History, Occasional Papers* (Leicester, 1954), no. 6.

22

The Land

The problems of drainage stemmed from the lack of gradient in the fenland.[1] Most of the fenland lies between 5 and 14 feet above O.D. This highest part of the fenland is not on the landward margin but the narrow silt ridge of the Townland, which lies only a short distance inland. Here the surface of the fen reaches about 12 feet above O.D. There is only a fall of a few feet to the coast, where the land is at approximately 8-10 feet above O.D. However, the average spring tide in the Wash reaches 14 feet above O.D., so that some form of coastal protection is essential. The lowest parts of the fen occur on the peats and heavy clays of the interior fenland. In Deeping Fen, for example, there is land which is only 5 feet above O.D. This means there is a negligible gradient for rivers and drains flowing through the fen to the Wash.

The Townland was the earliest part of the fen to be settled and drained. In each parish drains were cut which ran from the silt ridge to the Wash, and flowed by gravitation. These drains entered the Wash through sluice gates in the coastal banks (Fig. 4). In the middle of the eighteenth century they provided most of the Townland with an adequate drainage. Those in South Holland were amongst the least successful, for parts of the Marsh were as high or higher than the Townland, and there was an inadequate gradient. Nevertheless, the danger to the Townland and the Marsh came more from the sea than the drains. The Marsh was a result of accretion and reclamation in the Wash over a long period and so each enclosure required a further sea bank to be built. By the end of the eighteenth century the bulk of this reclamation had been achieved, and the last great enclosure took place in Holbeach and adjacent parishes in 1793, when 4500 acres were enclosed and embanked. The banks built and maintained by each parish were adequate to prevent all but the highest tide overflowing the fenland, and between 1770 and 1850 this occurred only once, in 1811. In that year a high spring tide with a following north-easterly breached the banks and much of Skirbeck and Kirton Hundreds were flooded.

[1] The most important description of the drainage of the Lincolnshire Fenland is to be found in W. H. Wheeler, *History of the Fens of South Lincolnshire* (Boston, 1896). Briefer descriptions can be found in H. C. Darby, *The Draining of the Fens* (Cambridge, 1940), and F. M. Tomes, 'The drainage of the Witham Fens', *Lincolnshire Magazine*, 2, (Lincoln, 1934).

The Agricultural Revolution in South Lincolnshire

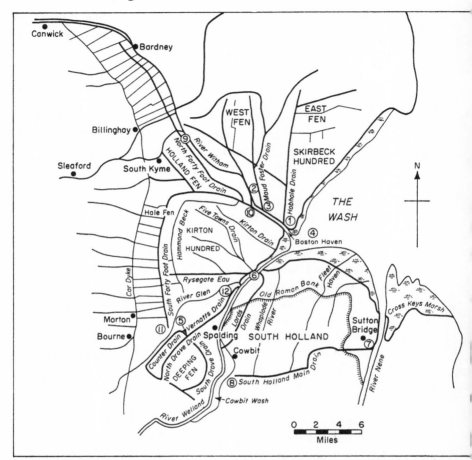

Fig. 4. The drainage of the fenland in the mid-nineteenth century (after W. H. Wheeler). 1, Hobhole Sluice; 2, Grand Sluice; 3, Maud Foster Sluice; 4, Clayhole; 5, Podehole; 6, Fosdyke Bridge; 7, Peters Point; 8, Peak Hill; 9, Chapel Hill; 10, Black Sluice; 11, Gutherham Gowt; 12, Vernatts Sluice.

Landward of the silt ridge the problems of fen drainage were much greater and were not finally solved until the late nineteenth century. Four rivers entered the fenland and flowed into the Wash: the Witham, the Welland, the Glen and the Nene. In their upper courses they had relatively steep gradients. Consequently, when they entered the fens they were unable to carry their load and

The Land

meandered across the plain depositing silt and flooding the surrounding countryside. To prevent this the Welland, the Glen and the Nene had been embanked by the mid-eighteenth century. This, whilst confining the rivers to a straighter course, did not prevent flooding. Silt was now deposited on the river bed and the rivers gradually rose above the surrounding land. The banks had to be constantly built up to prevent the rivers from overflowing.

The Wash was—and is—extremely shallow, and much of it was exposed as mud flat at low tide. Thus the rivers had to pursue a course from their *apparent* outfalls at Boston, Spalding and Sutton Bridge to an outfall proper some distance out in the Wash. The ebb tide repeatedly silted these courses up, and so instead of being able to cut a permanent lower reach, the rivers followed shifting, shallow courses at low tide. As long as these lower reaches were not cut down, the middle reaches in the fen could not be lowered. Consequently the gradient in the fenland remained small, and flooding inevitable.

The major rivers overflowed after heavy rain; but they were not the only source of floods. Many small rivers flowed on to the fen from the surrounding uplands, and their waters had to be collected in the drains and then conveyed into the rivers. Rainfall on to the fenland also had to be dealt with. The lack of gradient meant these waters had to be diverted into parish drains, and then into the rivers. Thus the fen drainers faced a number of difficulties. They had to maintain a deep course in the outfalls of the main rivers; to prevent the rivers meandering or silting in their middle reaches; to collect the 'highland'[1] waters into the drainage network; and to move the fen waters from the land into the rivers. By the end of the eighteenth century the main outlines of the present fen drainage system were already sketched in. The rivers were the main arteries of the drainage system. A number of major drains—the South Forty Foot, Hobhole Drain, Maud Foster Drain, Vernatts Drain and the South Holland Main Drain—pushed back from the rivers into the interior of the fen. Minor drains in each parish collected the fen and highland waters and fed them into the main drains. This was done by means of scoop wheels driven by windmills. The main drains fell into the outfalls of the rivers through sluice gates. These sluice gates

[1] The higher land surrounding the fen.

The Agricultural Revolution in South Lincolnshire

prevented the ebb tide in the Wash backing up the drains, and silting and flooding their lower courses.

In the remainder of this chapter the great fen drainage schemes of the later eighteenth century are outlined. Although these schemes have been described before, the extent to which they actually operated efficiently has received less attention. Because of the overriding importance of drainage efficiency for the development of agriculture in the fenland, the schemes are discussed in some detail. This is most conveniently done by Drainage Districts: the Witham and the Witham Fens; East, West and Wildmere Fens; the Black Sluice Level; Deeping Fen and finally South Holland.

The River Witham was the last of the four rivers to be embanked and in the middle of the eighteenth century it still flowed unconfined from Lincoln to Boston, flooding the surrounding fens periodically. In 1761, however, a Trust was established to make the Witham navigable and to improve the drainage. The area to benefit from this drainage was divided into six districts, and a drainage rate imposed on the occupiers of land. The General Commissioners raised a capital sum on mortgage and began the improvements in 1762. The scheme included: the cutting of a completely new course for the Witham between Chapel Hill and Boston: the deepening and embanking of the Witham between Lincoln and Chapel Hill; the construction of locks to improve the navigation; and the building of the Grand Sluice in Boston to prevent tidal waters entering the river and silting the bed.[1]

The Grand Sluice was completed in 1766, and at first the drainage of the Witham Fens seemed greatly improved. However, the drainage soon began to decline in efficiency. There were two reasons for this. The first was that insufficient money had been raised to complete the scheme. Whilst the new course from Chapel Hill to Boston had been successfully completed, the embanking and deepening of the river above that point had not been begun. Thus Holland Fen, which lay to the south-west of the new course, had been successfully drained; but the parishes between Chapel Hill and Lincoln which lay beside the unembanked and unscoured section of the river remained in 1776 'almost washed level as they were

[1] Wheeler, pp. 144, 152, 154.

The Land

before they were made, and therefore the money is almost all sunk'. Most of these parishes, encouraged by the successful drainage of Holland Fen, had taken out acts to enclose and drain their own fen. Their minor drains fed into the Witham by means of scoop wheels powered by windmills. But they were, of course, dependent on the Witham for their drainage, and the failure to deepen the Witham above Chapel Hill meant that it continued to flood.[1]

The second and more serious reason for the failure of the Witham scheme was the condition of the Grand Sluice. The ebb tide constantly deposited silt in the course of the Witham below the Grand Sluice—Boston Haven as it was called—and prevented an adequate scour being developed. The river between the sluice and its outfall at Clayhole remained a shifting, meandering course and thus prevented the reduction of the gradient higher up. In some years silt accumulated on the gates of the sluice and prevented them from being opened. This happened at the end of the summer of 1799, and the autumn rains could not be discharged into the Haven. The result was that the Witham overflowed along its whole length. This was a particularly bad year. But every year the autumn and winter flood waters spent their energy clearing away the tidal silt rather than cutting the outfall course lower, so that there was a steady decline in drainage efficiency throughout the Witham Fens. In 1807 a visiting engineer wrote that 'every year many thousand acres lie in a drowned state until it is too late to sow them with corn to advantage; this [the drainage] has seldom been affected, although attempted at a great expense by wind engines. These lands therefore can scarcely be called more than half-yearly lands.'[2] Flooding during the latter part of the Napoleonic Wars was particularly bad and there were renewed efforts to improve the drainage. Under the Witham Navigation Act of 1812 the outfall between the Grand

---

[1] S. G. S. Banks Stanhope; John Smith, *Report and opinion on the drainage of the River Witham* (1776); Wheeler, p. 183; L.A.O. 2 B N.L.; S.G.S./B.S./14.

[2] W. Jessop, *Report to the Commissioners of the Witham Navigation* (Newark, 1793); W. Chapman, *Observations on the Improvement of Boston Haven* (Boston, 1800); John Rennie, *Report on the Improving of the Witham Navigation* (Boston, 1802); *Report concerning the Improvement of Boston Haven* (Boston, 1800); *Report on the Proposed Improvement of the Drainage and Navigation of the River Witham* (Boston, 1803); *Report to the General Commissioners for Drainage and Navigation of the River Witham* (Boston, 1807); Wheeler, pp. 158–63.

The Agricultural Revolution in South Lincolnshire

Sluice and the Maud Foster Sluice was improved, and work was begun cutting a new channel between Lincoln and Bardney; but the lack of money together with labour troubles led to the scheme being suspended until 1825.[1]

To the east of the Witham lay the East, West and Wildmore Fens. These extensive fens lay unenclosed and undrained until 1801; and so poor was the state of drainage that Arthur Young was rowed over them when he visited the region in 1799. Part of the reason for the long delay in enclosure was the fact that all the villages in Skirbeck Hundred together with a number in Lindsey to the north all held rights of common there. An act to enclose and drain was passed in 1801 and the drainage works were completed by 1807. Before 1801 what little drainage these fens possessed was by way of the Steeping river which flowed eastwards into Wainfleet Haven. Rennie, who undertook the drainage, cut a catchwater running around the northern side of the fens, thus collecting the highland water from the Wolds. To drain the fens themselves two major drains were cut southwards to empty into Boston Haven. The Maud Foster drained Wildmore and West Fens, falling into Boston Haven below the Grand Sluice through the Maud Foster Sluice; the East Fen was drained by the Hobhole, which fell into Boston Haven through the Hobhole Sluice, also below the Grand Sluice. This drainage system proved to be one of the most effective installed in the area. Much of the fen was converted to arable, and at the end of the Napoleonic Wars the area was reported to be in an excellent state of drainage.[2]

Between Boston and Bourne lay an extensive area of fenland which in the mid-eighteenth century had little effective drainage. However, the Witham Act was almost immediately followed by an act to drain this Level and work was begun in 1765. An existing drain ran from Hale Fen to Boston. This was deepened and embanked, and a new sluice, the Black Sluice, built to fall into the Haven below the Grand Sluice. An entirely new drain, the South Forty Foot, was constructed from Gutherham Gowt near the River

[1] Wheeler, pp. 254–6, 273–90; J. S. Padley, *Fens and Floods of Mid-Lincolnshire* (Lincoln, 1882); S.G.S./Wilson.

[2] Arthur Young, *A General View of the Agriculture of the County of Lincoln* (London, 1799), p. 232; Wheeler, pp. 216, 223–7; A. Bower, *Statement as to the Drainage and Levels of the Fens North of Boston* (Boston, 1814).

The Land

Glen to join the old drain at Hale Fen. The construction of this drain prompted the adjacent parishes in Kesteven and Holland to enclose and drain their parochial fenlands, and by the beginning of the Napoleonic Wars only three parishes in the Black Sluice District remained unenclosed.[1]

As with the Witham drainage, the improvement was initially considerable, but by the end of the century a decline had set in. There were a number of reasons for this. The South Forty Foot, together with the old drain from Hale to Boston was over 21 miles long. Because of the very small gradient, there was a tendency for the drain to overflow. Whenever minor improvements were carried out, scouring inevitably began at Boston and worked backwards up the drain. The chronic shortage of money meant that the work was often discontinued before the whole length of the drain had been cleaned. In addition, the initial improvement in the drainage caused the peat surface to shrink, reducing the gradient further. This affected not only the South Forty Foot, but all the major drains. But the most important reason was the condition of the Haven. The silting of the Haven prevented the opening of the Black Sluice at the appropriate times, just as it made the timely operation of the Grand Sluice difficult. A partial improvement was made by introducing windmills to drive the scoop wheels which transferred water from the parish drains into the South Forty Foot. But this was not a complete solution. In 1815 John Rennie wrote that 'the drainage is in a very imperfect state...the injury the lands sustain by being so frequently flooded is highly detrimental, and renders their value comparatively small to what it would be were they effectively drained...'. The drainage was least effective at the extreme south of the drain, and during the war period it was reckoned that the fens of Bourne and Morton were flooded three years out of every four. Farther north the fenland (in Great Hale and Heckington) was flooded every winter from 1795 to 1801.[2]

[1] Wheeler pp. 254–6, 273–90.
[2] Jarvis, Golding and Hare, *Report as to any improvement that might be made in the Drainage of the Black Sluice by the Removal of Obstructions between the Haven and Wyberton Roads* (Boston, 1799); John Rennie, *Opinion on the More Effectual Mode of Improving the Drainage between Boston Haven and Bourne* (London, 1815); *Annals of Agriculture*, xxxix (London, 1803), 558; L.A.O. 2 Cragg 1/5.

The Agricultural Revolution in South Lincolnshire

The whole of the area between the Welland and the Glen was known as Deeping Fen, but it consisted of three distinct sections. A central block of 15,000 acres had been enclosed and given a rudimentary form of drainage in the seventeenth century by a group known as the Adventurers. To the north-east of this block lay undrained and unenclosed fen held in common by the parishes of Spalding and Pinchbeck, whilst to the south-west of the Taxable Lands, as the Adventurers' lands were called, the fen was held in common by a number of Kesteven villages, of which Market Deeping was the principal.[1]

At the beginning of the war period neither of the sections of common land had any form of drainage whilst the Taxable Lands were quite inadequately drained. There were two outlets for the waters of the Taxable Lands. Water was pumped into Vernatts Drain by scoop wheels and transferred to the Welland outfall above Spalding through Vernatts Sluice. Vernatts Drain ran through the common lands, but the Adventurers had virtually sole use of the drain. Secondly, there was Lords Drain. Part of the Spalding common fen was used as a reservoir for the waters of the Taxable Lands; from there the water was taken in a tunnel under the Welland into South Holland where Lords Drain ran north and fell into the Welland outfall. Thus the Taxable Lands were drained at the expense of both the Spalding and Pinchbeck commons and of part of South Holland.[2]

Ultimately the drainage of the whole of Deeping Fen depended on the condition of the Welland outfall. The river had been embanked below Spalding but above the town it remained a shifting and shallow stream. The tidal waters in the outfall were often too high on the gates of Vernatts Sluice for them to be opened, and flood waters consequently backed up the drain and overflowed the fen. The condition of the outfall had been the responsibility of the Adventurers, but they had neglected it and in 1794 an act was passed not only to improve the outfall but to take the responsibility from the Adventurers. The act required the outfall to be deepened and embanked as far as Wyberton Roads; but as so often with fen drainage schemes, only part of the provisions were carried out. The Welland was

[1] Wheeler, pp. 316, 320–5.   [2] S.G.S./S.R./5/37.

The Land

improved only as far as Fosdyke Bridge. In 1801 the common lands of Deeping Fen were finally enclosed and plans for its drainage made. The whole of the fenland now came under one authority, the Deeping Fen Trust. The use of Lords Drain was stopped, and a number of new drains were cut; the Counterdrain, the North Drove Drain, the South Drove Drain and the Cross Drain. Both the River Glen and Vernatts Drain were improved.[1]

These undertakings, together with the earlier work on the Welland, led to some improvement; but it was again only temporary. The interior drains were joined together at Podehole, and the waters thence carried to the Welland by Vernatts Drain. But Vernatts Sluice was at a higher level than Podehole, so that even with the help of fifty windmills it was difficult to move water out of the fen. The condition of the Welland continued to be poor, and the sluice gates of Vernatts Drain could still not be opened at will. It is not surprising then that the fen was described in 1818 as being 'almost in a lost state'.[2]

Between the Welland and the Nene lay South Holland. Here there had been fewer attempts to drain than elsewhere. The parishes ran from north to south, each including a section of Marsh, Townland and Fen. The fen sections were drained by small parochial drains which ran from the extreme south to empty into the Wash through sluice gates in the coastal banks. But both the Townland and Marsh sections were at a higher level than the fen ends. Thus, as George Maxwell succinctly put it, 'the fen ends have no drainage'. A small part of South Holland was drained by the Old Shire Drain (also the county boundary) which fell into the North Main Level Drain and thence into the Nene by Gunthorpe Sluice. This area was rather better drained than the rest.[3]

But the condition of the greater part of the South Holland Fens was disgraceful. In 1792 an act was obtained to improve the drainage and as a result an entirely new drain, the South Holland Main, was cut, running from west to east across the fen ends. It began at

[1] Wheeler, pp. 298, 325–7.
[2] J. Rennie, *On the Improvement of the Outfall of Vernatts Drain* (London, 1818); T. Pear, *Report on the Improvement of the Outfall of the River Welland* (Bourne, 1815); B. Bevan, *Report on the Improvement of the Navigation and Drainage of the Welland* (1812). [3] S.G.S./S.R./5/37.

The Agricultural Revolution in South Lincolnshire

Peakhill near the Welland, and fell into the Nene outfall at Peter's Point. Each of the parishes cut minor drains to deliver their waters into the Main Drain. But the attempt was far from successful. In 1813 John Rennie wrote that all the efforts to drain South Holland had been 'hitherto without success. An act...was passed in 1793 ...but after extensive works have been executed and large sums of money expended...the drainage is still found to be incomplete... and many thousands of acres of valuable land are during the winter and spring so flooded that their produce is of little comparative value.'

The reason for this poor drainage was ultimately, as with all the fen drainage schemes, the condition of the outfall. The Nene outfall had been greatly improved by Kinderley's cut in 1753, but it had decayed since that date. The cut did not carry far enough into the Wash to prevent the meandering and silting at low tide that all the fen rivers suffered from. In addition the sluice of the South Holland Main Drain had been placed too near the sea. The Nene waters overrode the sluice for much of the time, whilst the tides were rarely low enough for the sluice gates to be opened. Thus the flood waters in the Main Drain could not be released at the proper time, and flooding was inevitable.[1]

By the end of the Napoleonic Wars the major drains of the modern fen drainage system had all been cut; there had undoubtedly been a great improvement in fen drainage between 1750 and 1815. But it would be quite wrong to suppose that the system was efficient. The greater part of the Interior Fen remained subject to flooding if not every autumn, at least in three out of every four years. This was because, first, water could not be reliably moved from the minor drains to the major drains; and second, because of the poor condition of the outfalls. The solution to these two problems only came later in the nineteenth century and until the middle of the nineteenth century farming in the fenland remained dominated by the problem of flooding.

[1] J. Rennie, *A Plan for Completely Draining South Holland* (London, 1813); *Report and Plan for the Improvement of the Outfall to the Sea ... and for the Drainage of the North Level and South Holland* (London, 1814); J. Carter, *Observation on the Proposed Improvement of the Nene Outfall in reference to the Drainage of South Holland* (Spalding, 1822); Wheeler, p. 106.

CHAPTER II

PRICES AND PROGRESS

Not only agriculture, but the whole of the British economy began to grow more quickly in the later eighteenth century, and indeed agricultural progress was to a considerable extent dependent on changes which took place in transportation, banking and agricultural prices. When exactly these changes began has been the subject of some dispute but it was almost certainly in the second half of the eighteenth century. Much of this controversy is due to the lack of reliable series of statistics and the difficulties of interpreting them; this is as true of South Lincolnshire as it is for the country as a whole. Because there are no figures of agricultural output or productivity the growth of South Lincolnshire agriculture can only be traced indirectly. The available data are shown in Fig. 5, and Tables 1 and 2 (pp. 35 and 36). The progress of enclosure is one good index of development and Fig. 5c shows the accumulative area enclosed by Parliamentary means.[1] Before 1762 only a dozen acts had been passed, but in the following two decades 45 acts enclosing over 100,000 acres were implemented. In the 1780's there was a lull, as there was in the country as a whole, but the enclosure movement was renewed with vigour in the 1790's. In that decade 30 acts were passed, 15 in the 1800's and 22 between 1810 and 1820. By 1820 the enclosure of South Lincolnshire was virtually complete; in the preceding seventy years over 250,000 acres had been enclosed. There appear to have been two critical periods; the 1760's when enclosure first got under way, and the 1790's when it was renewed after a decade of inactivity.

The progress in road improvement does not show quite the same trend. Fig. 5e shows, for any year, the accumulative mileage which had been placed under the control of a Turnpike Trust. Here the critical period was between 1755 and 1765 when the majority of

---

[1] The graph is based on the enclosure awards for Kesteven and Holland: Kesteven County Council, Sleaford; Holland County Council, Boston.

The Agricultural Revolution in South Lincolnshire

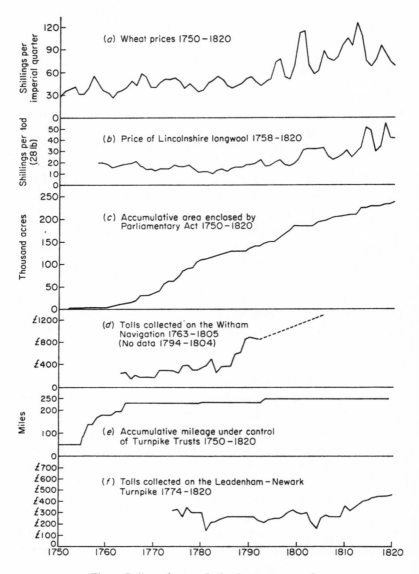

Fig. 5. Indices of economic development, 1750–1820.

Prices and Progress

the Turnpike Acts for the area were passed.[1] In contrast the timing of developments in water transport corresponded very closely with the trends in enclosure activity. The Witham and Black Sluice Acts (see above, pp. 26 and 28) in the 1760's contained provisions for the improvement of navigation as well as drainage. In 1781 Bourne and Spalding were linked by the deepening of Bourne Eau and the River Glen. The second critical period came in the 1790's when the Sleaford Navigation and the Grantham Canal were constructed, and the River Welland and the Fossdyke, for long theoretically navigable, were improved. The tolls collected on the three principal navigations—the river Witham, Sleaford Navigation, and Grantham Canal—illustrate the growth of traffic. On the Witham (Fig. 5d), there was a slow increase in the 20 years after the opening of the Navigation, but from 1783 until 1805, when the series ends, the tolls collected increased more rapidly.[2] On the Grantham Canal and the Sleaford Navigation there are only data for selected years, but it suggests a steady increase of trade.

TABLE 1. *Tolls on the Sleaford and Grantham waterways*

The Sleaford Navigation	The Grantham Canal	
1795 £498	1798 £4381	1801 £5150
1806 £562	1799 £1500	1803 £3981
1824 £1010	1800 £4031	1804 £5028

Sources: J. Creasey, *Sketches illustrative of the Topography and History of New and Old Sleaford* (Sleaford, 1825), p. 74; E. Turnor, *Collections for a History of the Town and Soke of Grantham* (London, 1806), p. xiii.

A final index of growth is rent. Whilst rent is a production cost for the farmer and gross income for the landlord, rental trends are often a reasonable index of agricultural prosperity. Unfortunately there are no continuous series available for South Lincolnshire before 1812 nor are there for England as a whole, so some isolated examples of rental increases must suffice. It seems that rents rose slowly in

[1] The graph is based on information contained in the Appendix of *The Report of the Commissioners for Inquiring into the State of the Roads in England and Wales* (London, 1840), pp. 259–74.
[2] S.G.S./B.S./15/2/7.

the greater part of the eighteenth century, and then more rapidly in the war period. One writer estimated that the average rent per acre in England doubled between 1776 and 1812, whilst another estimate suggests a doubling of rent between 1790 and 1815.[1] In South Lincolnshire only isolated instances of rental increases can be quoted. Thus in 1806 Lord Boston raised his Moulton estate 21 per cent; in 1801 the Earl of Bristol raised his estate 25 per cent, and increases on the Heathcote estate in 1813 varied from 20 to 35 per cent.[2] Table 2, below, shows the total rental value of Canwick, and the rent per acre of land at Castle Bytham and Threekingham at various times in the eighteenth century. This suggests that rent increased more rapidly during the war period than earlier in the century.

TABLE 2. *Rentals in selected parishes, 1727–1814*

Canwick		Castle Bytham	
1727	£350	1700	3s. 4d. per acre
1760	£730	1778	4s. 6d. per acre
1790	£1380	Threekingham	
1802	£1782	1795	20s. 0d. per acre
1812	£3200	1814	36s. 0d. per acre

Sources: L.A.O. Brace 14/20/13; *The Victoria History of the County of Lincoln* (London, 1906), II, 345, 351.

Before considering the regional pattern of rental increase a number of qualifications should be made about the nature of rent. Landlords differed considerably in their policies on rent, particularly in the inflationary conditions of the war. Many old-established landlords only valued their estates once a generation, and were not prepared to increase rents simply because there were high grain prices. In fact they 'estimate their reputation and character too high to allow them to squeeze and suppress those whom providence

[1] T. S. Ashton, *An Economic History of England; the Eighteenth Century* (London, 1955), pp. 45–6; F. M. L. Thompson, 'The English land market in the nineteenth century', *Oxford Economic Papers*, IX, no. 3 (Oxford, 1957), 289; R. J. Thompson, 'An inquiry into rent in the nineteenth century', *Journal of the Royal Statistical Society*, CXX (1907), 587.

[2] L.A.O. 2 Cragg 1/6, Boston 1/1/30, 2 Anc 6/27, Jarvis 8.

Prices and Progress

has placed below them'. Other landlords were less paternal and in the 1800's—particularly in the fenland—extraordinary rents were asked. Indeed there were 'two sorts of scales of rent...for the same quality of land...those paid to the reputable landlord; and those paid to the jobbers or dealers of land'. There were other differences in policy. Thus at Doddington a land agent recommended Colonel Jarvis to raise his rent in order to force the tenants to undertake improvements. On the other hand tenants of the Earl of Bristol who had made improvements at their own cost were compensated by a reduced rental increase when the estate was revalued.[1] Clearly then rental increases did not reflect economic conditions alone. However, a comparison of the Property Tax returns for each parish in 1806 and 1815 shows that rent was increasing most in the areas which had been recently enclosed and drained.[2] In many cases in these regions, the fen and the Heath, land was being farmed for the first time. The lowest increases came on the clay lands of south-east and west Kesteven. On the Heath there were some staggering rental increases after enclosure. A parish near Sleaford which had been valued at £223 in 1771 was rented at £3000 in 1824, and there were other equally striking examples.[3]

The information presented above, sporadic as it is, leaves no doubt that there was a marked acceleration in economic activity in South Lincolnshire after 1750. The reasons for this must be briefly discussed. The major cause seems to have been the more favourable agricultural prices of the second half of the eighteenth century (Fig. 5 *a* and *b*, p. 34). It was not, however, any sudden increase in prices which led to the changes. On the contrary, as a South Lincolnshire farmer wrote, 'as farmers are accustomed to see great and sudden fluctuations in the marketable value of the different products of their industry, without much affecting their average value from year to year, such a crisis in rural affairs must manifest

[1] *Farmer's Magazine*, VI (Edinburgh, 1805), 511; VII (Edinburgh, 1806), 120–1; L.A.O. Jarvis 8, 2 Cragg 1/6.
[2] The use of the Property and Income Tax assessments of the nineteenth century is examined in D. B. Grigg, 'Changing Regional Values during the Agricultural Revolution in South Lincolnshire', *Transactions and Papers of the Institute of British Geographers*, no. 30 (London, 1962), pp. 91–103.
[3] Creasey, p. 369.

itself by effects sufficiently steady, continued and uniform before this experienced and wary class of men will be persuaded to alter materially their general course of management'.[1] After 1760 these conditions were fulfilled. Between 1730 and 1750 agricultural prices had been unfavourable, and indeed one writer has described this period as an agricultural depression,[2] but after 1760 there was a steady improvement in wheat prices. The average annual price of wheat between 1760 and 1794 was one-third above that of 1720-59.[3] Later wheat prices rose prodigiously and when in July, 1796, 200s. was asked for a quarter at Stamford, 'a great tumult arose which caused a large body of special constables to be sworn in to attend the following markets'.[4] From 1794 to the end of the war bad harvests, the fear of invasion and currency inflation combined to keep grain prices at extraordinary heights. Although this benefited farmers and landlords, it was hard on the rest of the community. In South Lincolnshire rice, barley and oats were tried as substitutes for wheaten flour; as a last resort millers were threatened. A series of notes to a Market Deeping miller culminated in the following: 'If you dont fall your flower [sic] next Saturday to 3/6 a stone we will pull your mill down over your head and take your flour away. We will Dragg you about the market place on Saturday next.'[5]

Important as wheat prices were to farmers, wool prices were perhaps more significant to South Lincolnshire farmers. It was not until the end of the American War that they shared in the upward movement and before this low prices had caused widespread distress (Fig. 5b, p. 34). 'Unless some expedient is hit upon', wrote a Holbeach land agent in 1779, 'to keep up the price of wool, our county is ruined.' Two years later he wrote that 'nothing but peace can prevent the impending ruin'.[6] In fact wool prices did rise after 1783, and during the Napoleonic Wars were well above any preceding level. Livestock prices were also rising in the second half of the eighteenth century; in particular mutton was rising at a greater rate

---
[1] *Annals*, XLIII (1805), 431-2.
[2] G. Mingay, 'The agricultural depression, 1730-1750', *Economic History Review*, VIII, no. iii (1956), 323-38.
[3] A. J. Youngson, *Possibilities of Economic Progress* (Cambridge, 1959), p. 122.
[4] G. Burton, *Chronology of Stamford* (London, 1846), p. 116.
[5] P.R.O./H.O./42/53/488.   [6] L.A.O./T.Y.R. 4/1/100.

Prices and Progress

than wool, and it was this which stimulated farmers to cross their Longwool sheep, an animal bred solely for wool, with the Leicester; the cross-breed gave mutton and wool.[1]

'Nothing can tend so much to the improvement of land', wrote a Branston farmer in 1816, 'as the continued exercise of capital upon it...'.[2] In the preceding sixty years farmers and landlords had taken this advice. In the 1820's outsiders commented on the considerable investment in improvements in South Lincolnshire, not only in matters directly concerned with agriculture but also in the improvement of transport facilities. There is no means of knowing the total expenditure on improvements in this period, but some examples of the outlays undertaken will give an idea of the magnitude of investment. Enclosure was probably the largest single item. Costs varied a great deal. In North Lincolnshire it has been estimated that the average expenditure was 25s. per acre. Young quotes the cost of enclosing six parishes in Kesteven as just under 17s. an acre. In the fenland costs were certainly much higher, for enclosure acts often included provisions for drainage undertakings as well. The combined cost of enclosing and draining East, West and Wildmore Fens was £400,000, Holland Fen £50,600. The expenditure on drainage schemes was enormous. It took £17,985 to cut the South Holland Main Drain, and £53,650 to carry out only part of the provisions of the 1762 Witham Act. The improvements to roads and waterways was equally expensive. The Grantham Canal cost £100,000 whilst Fosdyke Bridge, built in 1811 across the Lower Welland, necessitated an outlay of at least £20,000.[3]

The means of financing these projects varied, but most of the money was raised locally. Turnpikes and canals were financed by a mortgage on the tolls, and drainage schemes by raising a loan, the interest being paid by a drainage rate on the proprietors.[4] Enclosure was financed entirely by the proprietors who partly recouped their expenditure by increasing the rents. The spread of country banks

---

[1] W. Marrat, *A History of Lincolnshire* (Boston, 1814), I, part 2, 96.
[2] C. White, *A Short and Plain Letter on Agricultural Depression* (London, 1816), p. 17.
[3] *J.R.A.S.E.* IV (1843), 291; Wheeler, p. 156; Young, *General View*, pp. 79, 87, 240, 272; W. E. Tate, 'The cost of Parliamentary enclosure in England', *Economic History Review*, V, no. ii (1952), 262.   [4] Pressnell, p. 381.

undoubtedly assisted the raising of capital, but only towards the end of the century. In 1754 there were only 12 country banks in the whole of England but this had grown to 230 in 1797 and to 721 in 1810. South Lincolnshire was well provided with them and by 1800 there were four at Stamford and at least one at Bourne, Sleaford, Grantham, Boston and Lincoln.[1] The banks seem to have been willing to extend credit to both landlords and farmers, and at the end of the war period a local farmer wrote 'that nothing had contributed more to the advancement of agriculture than the assistance country bankers furnished to the occupiers of land'.[2] The banks advanced credit not only to farmers but to drainage trusts, canal companies and enclosure commissioners. Peacock and Handley, a Sleaford bank, made loans to the Sleaford Navigation, the Deeping Fen enclosure commissioners and also for the improvement of the River Welland.[3] Indeed there were some who thought the local banks too free with their credit, and attributed the inflationary price conditions to their activities. 'The increase of country banks', wrote a Leadenham landlord, 'who supply jobbers of every denomination from the corn factor to the pig jobber with their bills... I have asked a great many farmers their opinions of the present high prices of corn, and their rustic answer was, "We shall have no good doings, Sir, whilst there is so much paper money about".'[4]

But the establishment of country banks came late in the period of development. How were improvements financed before this? There seem to be two possibilities. The first is simply that both landlords and farmers were prepared to raise the proportion of income they ploughed back into the land in the hope that the more favourable price conditions would reward them. Alternatively it may have been that in the later eighteenth century agricultural prices rose more than production costs, thus leaving an increasing profit margin that could be invested. Unfortunately this is difficult to confirm as there is little reliable data on production costs. Labour was one of the most important items, and wages certainly rose after 1783. Much of the newly farmed land on the Heath and the fen was distant from

[1] H. Porter, 'Old Private Banks of Southern Lincolnshire', *Lincolnshire Magazine*, III, no. 2 (1935); 'Lincolnshire Private Bankers', *Lincolnshire Magazine*, vol. 5, no. 3 (1937).
[2] White, p. 15.   [3] Pressnell, pp. 354, 355, 392.   [4] P.R.O./H.O. 42/53.

Prices and Progress

settlement, and there were acute labour shortages. In addition many labourers left the farms in Holland to work on the better paid drainage schemes, so that by 1790 the fen harvest was dependent on migrant Irish labour.[1] One observer at the end of the war period considered that total production costs had doubled between 1790 and 1815. Certainly at the end of the war period wages in Lincolnshire were among the highest in the country.[2]

Unfortunately, without a great deal more information than is available it is impossible to trace the mechanism of investment in agriculture at this period. However, two conclusions can be drawn. First, the bulk of the investment was undertaken by landlords rather than tenants; they were responsible for both the costs of enclosure and drainage. For reasons which will be discussed in the next chapter tenant investment was very meagre in this period. Secondly, most of the investment went into projects which expanded the agricultural area—such as enclosure and drainage—rather than in methods of increasing output per acre.

Much of the agricultural development in the area would have been impossible without the great improvements in the roads and waterways. And certainly in the middle of the eighteenth century there was a great need for improvement, for the roads were in a deplorable condition. This, however, was to some extent ameliorated by the spread of Turnpike Trusts, which became responsible for maintenance and were allowed to charge tolls on the traffic. An act of Parliament was required for the establishment of a Trust, and this enables the progress of road improvement to be traced. Fig. 6 shows the turnpike roads of the area and the date of the first act effecting them. Before 1750 only about 50 miles of road had come under the control of Trusts, mainly on the Great North Road between Stamford and Newark. But between 1755 and 1765 acts effecting a further 180 miles were passed, over half the total mileage eventually placed under Trust control in South Lincolnshire (see Fig. 5e, p. 34). The acts passed *after* 1765 mainly affected the fenland, where there were particular difficulties in building roads. Not

---

[1] *Annals*, XIX (1793), 188; *Farmer's Magazine*, VI (1805), 511; IX (1808), 409.
[2] *B.P.P.* III (1813–14), 48; V (1814), 1035; VI (1824), 401.

The Agricultural Revolution in South Lincolnshire

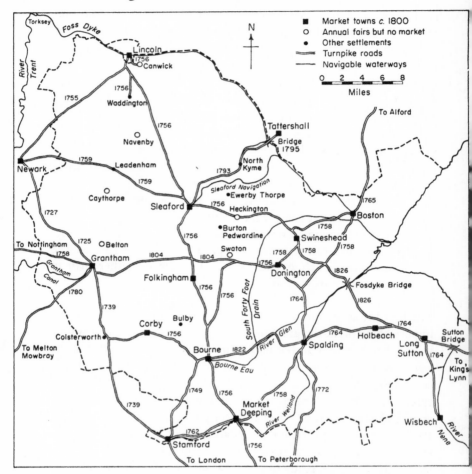

Fig. 6. Turnpike roads and navigable waterways.

only was there a lack of suitable local materials for construction, but much of the Interior Fen was flooded in winter rendering the roads impassable. However, in 1793 the Sleaford to Tattershall Turnpike Trust was established, and built the first bridge over the Witham between Lincoln and Boston. Eleven years passed before the Grantham–Bridgend turnpike finally linked Grantham and Boston. In 1811 Fosdyke Bridge was built across the Welland below

Prices and Progress

Spalding; but it was not until the end of the 1820's that the Nene was bridged below Wisbech, providing a direct route between Boston and Kings Lynn.

The fact that a road had been placed under a Turnpike Trust did not by any means guarantee improvement in its condition. Until the 1820's, when MacAdam worked in the area and his ideas gained ground, turnpikes were often far from adequate.[1] The prohibitive costs of transporting suitable building materials, meant that local materials were generally used. The fen roads for example were made from sand or silt 'digged from the subsoil of the adjacent fields; and this mode of repairing them occasions a heavy expence to the traveller'. Only the better fen roads had a gravel base. In Kesteven the poorest roads were found in the clay areas, where the lack of good building materials was particularly serious, for the clays became almost impassable in wet weather. Arthur Young's conclusion that the roads of Lincolnshire were 'below par' was supported by local writers.[2] Before the enclosure of Holland Fen in 1770 travellers from Sleaford to Boston had to take a guide; 50 years later the same route 'was so heavy even in the summer season that wheel travelling was almost impracticable'. Until 1823 the road from Spalding to Bourne was so flooded in wet weather that the journey had to be made by boat, whilst the Townland roads, although free from flooding, were impassable for three-quarters of the year.

It might be thought then that the establishment of turnpike trusts had little effect on road conditions. One authority considers that their establishment did little to reduce transport costs, and in parts of South Lincolnshire there seems to have been little increase in the freight carried.[3] Fig. 5f, p. 34, shows the tolls collected between Leadenham and Newark.[4] Until the end of the Napoleonic Wars there was no marked upward trend; yet between 1825 and 1845 (when McAdam was at work) tolls were twice the annual average

---

[1] C. Brears, *Lincolnshire in the 17th and 18th centuries* (London, 1940), p. 156.
[2] *Farmer's Magazine*, I (1800), 394; S.G.S./B.S./11 A; B.P.P., v (1818), 211–12; Young, *General View*, p. 405; *Annals*, XXXV (1800), 390.
[3] W. T. Jackman, *The Development of Transportation in Modern England*, I (London, 1916), 348–9.
[4] A. Cossons, 'The Turnpike Roads of Nottinghamshire', *Historical Association Leaflet*, no. 97 (1934), p. 33.

of the last quarter of the eighteenth century. Yet other evidence shows that in some parts of South Lincolnshire at least, the establishment of turnpikes had remarkable consequences. This was particularly true of the Sleaford–Tattershall turnpike of 1793, which ran through a number of formerly isolated fen villages. Before its construction '...east of Anwick, all intercourse nearly ceased with the autumn rains'. North Kyme had been 'nearly inaccessible by land in winter', whilst the improvement in the road greatly increased the value of land around Billinghay. Nor was road improvement confined solely to those roads which came under turnpike trusts. Two small villages near Sleaford had their roads remade with a gravel base towards the end of the Napoleonic Wars. Before this the inhabitants of Burton Pedwardine had been 'shut up from almost all society for more than half their days', whilst at Ewerby Thorpe it 'had made such a change as seldom experienced when but a few years ago the roads hereabouts were hardly passable'.[1]

Important as the roads may have been it was certainly the improvement of waterways which had the greatest impact on South Lincolnshire farming. Some of the rivers—the lower Witham, the Welland and the Glen—had been navigable for some time prior to the eighteenth century, whilst the Fossdyke—which ran from Lincoln to the Trent at Torksey—had been cut in Roman times. But in the middle of the eighteenth century these waterways were in a bad condition and carried little traffic. However, in 1741 Richard Ellison leased the Fossdyke from Lincoln Corporation and undertook to improve the navigation. This was the first in a series of improvements. Between 1760 and 1770 both the Witham and the South Forty Foot Drain were made navigable, in 1781 the Glen and Bourne Eau were scoured out to link Bourne with Spalding, and in the 1790's the Grantham Canal and Sleaford Navigation were constructed. The former gave Grantham a water link with the Trent, the latter joined Sleaford with the lower Witham. By 1800 all the major market towns in South Lincolnshire (see Fig. 6, p. 42) had a waterway to either the North Sea or the Trent.

Nevertheless, the waterways were not without their troubles. The

[1] L.A.O. Cragg 1/1 (Billinghay, North Kyme); Creasey, p. 181.

Prices and Progress

Witham, the main artery of trade in the area, suffered a number of reverses, and Messrs Keyworth and Fowler who had leased the tolls had several complaints to make to the Commissioners in 1806. In winter the river often flooded the towpath and prevented horses hauling the boats whilst in the summer there was often too little water in the river for passage of vessels. In the dry summers of 1797 and 1798 there was hardly any water in the river at Washingborough whilst in 1801 and 1803 part of the course was emptied to enable the drainage to be improved. Nevertheless, trade on the Witham was considerable and it grew steadily from 1783 to 1805 (see Fig. 5 d, p. 34). On the Welland and the Fossdyke the improvements to the navigation were less effective and indeed trade on the Welland seems to have declined towards the end of the century.[1]

The improvements made to transport by water undoubtedly had a considerable impact on farming. To begin with it made it easier and cheaper for farmers to send their grain to the local markets (see Fig. 6, p. 42). Indeed in some cases the only way was by water. 'What are called roads at Chapel Hill', wrote John Cragg, 'are in winter not passable, and if it were not for the navigable rivers passing by the place the inhabitants would be miserable.' At South Kyme, the whole of the parish's grain was moved to market by water after the opening of the Sleaford Navigation. The same was true of the remote hamlet of Harts Ground, whilst John Cartwright, a Brothertoft farmer whose land lay on the banks of the Witham, sent all his woad, corn and sheep to Boston by water. In the fenland the numerous drains provided easy access not only for barges carrying bulk cargoes, but even to farmers' wives who rowed to market in small boats called *schoutts*.[2]

Outside the fenland there were of course fewer waterways; nevertheless the Grantham Canal, the Fossdyke and the Sleaford Navigation were of great importance for they allowed the cheaper import of various goods which farmers needed. Coal and timber came into South Lincolnshire by the Trent and then canal. But of more direct concern to farmers was the movement of fertilizers and

---

[1] S.G.S./B.S. 11A; B.S. 1.
[2] L.A.O. Cragg 1/1 (South Kyme, Hart's Ground, Chapel Hill); S.G.S./B.S./14; *The Lincolnshire Cabinet and Intelligencer* (Lincoln, 1827).

manures. Most waterways carried fertilizers at a greatly reduced charge; an exception to this was the Witham, where the full toll was exacted. John Cartwright was particularly indignant at this and wrote a number of letters to the Witham Commissioners, without much success. The reason for the toll—or rather for the failure to reduce it—was that when the Witham Act had been framed 'none ever dreamed ...that the fat fen land would ever stand in need of manures'.[1]

But the most important benefit of the new waterways was to link the South Lincolnshire farmer more easily with the rapidly expanding markets of the industrial north and midlands, and particularly with London. The Grantham Canal and the Fossdyke linked two major towns with the Trent, and thus with the West Riding, Nottingham and the West Midlands. Lincolnshire malt found its way over the Pennines to Lancashire, and some went even farther afield, for Newark beers were exported to St Petersburg. Both wool and livestock travelled by water to the West Riding, and in the nineteenth century Sheffield, Wakefield and Manchester were getting much of their meat supply from the markets of Boston and Lincoln. Grain too moved westwards to the Trent. In the first decade of the nineteenth century over 11,000 quarters of corn a year were moving west from Grantham alone. But the main direction of the grain trade seems to have been eastwards. All the major South Lincolnshire market towns except Grantham had water connexions with either Boston or Spalding; and from these two ports, although predominantly from Boston, grain was carried coastwise, especially to London. In 1811 over a third of all the oats imported into London for stables came from Boston. The growth of this trade during the war period is shown in Table 5, p. 71. For much of the first part of the nineteenth century Boston was one of the three leading corn markets in the country. Although northern towns were drawing meat supplies from the area, probably the bulk of the livestock were still driven south by hoof to Smithfield; wool on the other hand went almost entirely to the West Riding, and there were close connexions between South Lincolnshire farmers and the Yorkshire textile area.[2]

---

[1] S.G.S./B.S./11 A.
[2] *Stamford Mercury*, 12 Feb. 1830; L.A.O. Cragg 2/32/1; *Annals*, XXXI (1798), 201; Turnor, p. xiii; P. Thompson, *Collections for a History of Boston* (London, 1820), pp. 376–7.

CHAPTER III

FARMING METHODS AND PRODUCTIVITY

To nineteenth century writers the 'agricultural revolution' was a relatively simple process. They considered that the most important of the new methods was the introduction of improved rotations containing turnips and temporary grasses. These techniques were thought to have developed in Norfolk and then spread to the rest of the country. Because it was difficult to grow turnips on the fallow of open field farms, enclosure was believed to be the key to any form of progress. As the later eighteenth century was the period of greatest enclosure activity, it was thus thought to have been the period of most rapid technical change. In recent years these views have had to be modified. To begin with it is now recognized that some improvements were possible whilst land remained in open field; by special arrangement between farmers and cottagers for example, turnips could be grown on the fallow. More important, it is clear that whilst enclosure made the introduction of new methods a great deal more easy, it did not necessitate improvement. Secondly, there is now considerable evidence that many of the techniques of the New Husbandry were being practised in the seventeenth century, not only in Norfolk but in other parts of eastern England. Consequently there is a growing body of opinion that the critical period of technological change was the early, not the late, eighteenth century. Thirdly, it should be remembered that the new methods affected primarily arable farming; consequently the Norfolk system, although it may have spread throughout the arable area of eastern England, had little impact on the pastoral farms of the Midlands and the west. Even within the predominantly arable regions of eastern and southern England the new methods were not everywhere adopted. Because the turnip could not be easily grown on heavy soils, the clay regions felt little of the first phase of the agricultural revolution; most progress was made on the 'light' lands, particularly

The Agricultural Revolution in South Lincolnshire

the limestone regions which stretched from the Yorkshire Wolds, through Lincolnshire and then south-westwards to Hampshire and Dorset.[1]

Before discussing the state of farming in South Lincolnshire during the Napoleonic Wars, it would be as well to consider what exactly the new methods were. It should be noticed straight away that the introduction of farm machinery played little part in the agricultural revolution. Only the drill was of importance, although in some regions the introduction of improved plough designs assisted the more thorough cultivation which was general in the New Husbandry. Although a number of South Lincolnshire farmers were experimenting with other types of machinery during the wars, their adoption did not come until much later. There was not then any marked decline in labour needs as a result of the introduction of machinery. On the contrary most of the new methods required an augmented labour force. The land was tilled more frequently, and turnips in particular required a great deal of attention; the New Husbandry, which by Victorian times had evolved into the mixed farming of the 'Golden Age', was in fact dependent on an expanded and cheap supply of labour. This, until the 1840's at any rate, was probably adequately provided by the increasing rural population.

The central feature of the new methods was the integration of livestock and arable farming. It is true that both crops and stock had been raised on open field farms, but the two systems were only incidentally interlocked. The turnip was the key to integration. The bare fallow could then be utilized, providing feed for livestock whilst still resting the land from successive grain crops. Sheep could be folded on the crop, their feet consolidating the lighter soils, as well as manuring the land for the succeeding wheat crop. In addition the turnip, being sown in rows, allowed cultivation during its growth, thus serving as a 'cleaning' crop. The Norfolk four course also included a year under temporary grasses, usually clover, a

---

[1] E. Kerridge, 'Turnip husbandry in High Suffolk', *Economic History Review*, III, no. iii (1956), 390–2; R. Parker, 'Coke of Norfolk and the Agricultural Revolution', *Economic History Review*, VIII, no. iii (1955), 156–66; J. H. Plumb, 'Sir Robert Walpole and Norfolk Husbandry', *Economic History Review*, V, no. i (1952), 86–9.

Farming Methods and Productivity

feature of the system which has perhaps been neglected, for as well as providing grazing the crop restored the nitrogen content of the soil after its depletion by grain crops. These crops, generally wheat or barley, were cash crops, whilst their straw was used in the preparation of farmyard manure. At the end of the eighteenth century cattle were increasingly being fed in stalls on purchased oil-cake, as well as on fodder crops grown on the farm. This gave a greatly enriched farmyard manure; in addition some farmers were using artificial fertilizers, of which lime and crushed bones were the most important.

These were perhaps the most important of the new methods, but there were others. In the clay regions the growth of turnips and the use of manures was difficult where the land was not adequately drained. Most farmers still relied on some form of crude surface drainage, but experiments were under way with 'hollow' drainage. Enclosure, which greatly reduced the intermixture of holdings and led to the fencing of fields, made selective breeding easier. In the eighteenth century many landlords were experimenting in producing new animals, so that there was a slow improvement in the quantity and quality of livestock products.

Although there are isolated examples of improvement on the open fields before enclosure, the persistence of common land was more usually a barrier to improved farming. Neither turnips nor clover could be grown and underdrainage was difficult while holdings remained intermixed. Selective breeding was impossible whilst the fields were grazed in common and holdings remained unfenced. So before the state of South Lincolnshire farming is discussed, the progress of enclosure in the area must be examined.

There were few parishes in South Lincoln which had not experienced some enclosure, however small, before the era of Parliamentary enclosure. There were, however, three regions where the open fields had been eliminated long before the middle of the eighteenth century[1] (see Fig. 7, p. 51). In west Kesteven there was a zone of

---

[1] The map is based on: L.A.O. 1/1 (John Cragg's Topographical Dictionary); D. R. Mills, 'Population and Settlement in Kesteven, 1775–1885', M.A. thesis, University of Nottingham (1957), vol. II; W. V. R. King Fane, *A Schedule of Kesteven*

The Agricultural Revolution in South Lincolnshire

old enclosed parishes stretching from Lincoln to Grantham across the Lias clay plain. In the extreme west of this region a few open-field parishes still remained in 1750, but by the beginning of the Napoleonic Wars only Long Bennington remained unenclosed. South and east of Sleaford lay a second well-defined region of old enclosure, corresponding very closely with the boulder clay cover of the limestone Heath. Enclosure had begun here in the fourteenth century, and by 1750 only a few peripheral parishes remained unenclosed. The third region of old enclosure was in Holland. Here the open fields had either never existed or had been enclosed at a very early date.[1] Nevertheless, land was still held in common throughout the fenland. The earliest enclosure had been on the silt ridge of the Townlands, where by the end of the eighteenth century very little common land remained. In the Interior Fens and on the marsh, however, the bulk of the land still remained in common in 1790. There were two noticeable exceptions to this. In Deeping Fen a central block had been enclosed by the Adventurers in the seventeenth century, whilst there had also been piecemeal enclosure at an early date in the interior fenland of South Holland. All three regions of old enclosure were in areas of heavy clays, which at the end of the eighteenth century were mainly under grass. It seems likely that they owed their early enclosure to the fact that they were much more suited to pastoral than arable farming, a circumstance which has been observed in other parts of England.[2]

In spite of this long history of enclosure there were still one hundred parishes with open fields in Kesteven in 1750, and many thousand acres of common land in Holland. But the Parliamentary enclosure movement in South Lincolnshire proceeded with such vigour that by 1790 only forty parishes remained unenclosed, whilst by the end of the wars only Stamford retained its open fields. By 1815 most of the common land had also been enclosed; only some

---

Enclosure Awards; Holland Enclosure Awards, The Muniment Room, County Hall, Boston; J. Thirsk, *English Peasant Farming* (London, 1957), p. 239.

[1] Thirsk, p. 14.

[2] For example in Leicestershire and Nottinghamshire: H. G. Hunt, 'The chronology of Parliamentary enclosure in Leicestershire', *Economic History Review*, x, no. ii (1957), 270; J. D. Chambers, *Nottinghamshire in the Eighteenth Century* (London, 1932), p. 150.

Farming Methods and Productivity

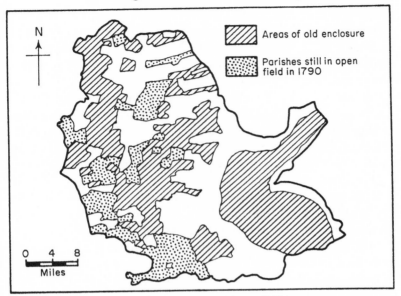

Fig. 7. Areas of old enclosure and open field in 1790.

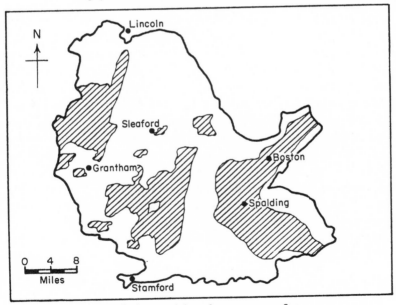

Fig. 8. Areas mainly under grass, 1795–1815.

The Agricultural Revolution in South Lincolnshire

6000 acres remained unenclosed in the whole area. The surviving open field parishes of 1790 lay with only a few exceptions in two regions (Fig. 7). On the limestone Heath there were a number of parishes near Sleaford, Grantham and Stamford which remained unenclosed until the high grain prices of the 1800's. This was doubtless due to the poor soil, which made cultivation unprofitable under normal price conditions. Around Market Deeping there was a second group of open-field parishes which had survived for quite different reasons. Each of these parishes had a large number of freeholders (see Fig. 10, p. 73), and as enclosure required the consent of a majority of the landowners, this subdivision may well have delayed enclosure. Furthermore, all the village open fields lay on the excellent gravel soils of the fen edge and so crop yields were above the average, removing some of the incentive to enclose. Lastly, all the farmers had unstinted grazing rights in Deeping Fen, and there was thus less pressure on their grazing resources.[1]

In Kesteven common land was invariably enclosed at the same time as the open fields; but in eighteenth-century Holland common land alone remained to be enclosed. By 1790 most of the Kesteven fenland had been enclosed, and this was broadly true of the fenland of Kirton Hundred in Holland. The parishes of Kirton had until 1770 rights of common in Holland Fen, in addition to the fen contiguous to each settlement. In 1770 Holland Fen was enclosed and the land allocated to each parish according to the number of grazing rights each parish had held when the fen had been intercommoned. Separate acts were passed to further divide the land pertaining to each parish. This was largely achieved by 1790. Intercommoning was also practised to the north of Holland Fen and the River Witham, for all the parishes in Skirbeck Hundred—with the exception of Wrangle—had grazing rights in East, West and Wildmore Fens. They shared these commons with a number of Lindsey parishes, and it was no doubt this extreme subdivision of rights which delayed the enclosure of these Fens so long.[2] It was not until 1801 that a general act to enclose and drain them was passed, whilst

---

[1] Thomas Stone, *A General View of the Agriculture of the County of Lincoln* (London, 1794), p. 22.
[2] Stone, *General View*, pp. 18–20.

Farming Methods and Productivity

the separate acts allocating land to each parish with grazing rights followed at various later dates. In a few cases the parish awards were not finally carried out until 1845.

South of Kirton the pattern of enclosure was more complicated. The land between the rivers Welland and Glen was all generally referred to as Deeping Fen, but the central portion had been enclosed in the seventeenth century; to the north-east of these Taxable Lands the common fen remaining was shared between Pinchbeck and Spalding, whilst to the south-west the unenclosed fen was intercommoned by a number of Kesteven parishes, the principal of which was Market Deeping. Both areas of common fen were enclosed in 1801. In South Holland the interior fenlands had experienced more piecemeal early enclosure than in Kirton and Skirbeck, where most of the enclosure was undertaken in the Parliamentary era.

As might be expected, the standard of farming on the open fields of South Lincolnshire was low, and examples of the type of improvements being undertaken which have been noticed in other parts of England were rare. In the few clay open-field parishes drainage, if attempted at all, was simply the ridge and furrow of surface drainage. Crop rotations rarely varied from the traditional (1) fallow, (2) wheat, (3) beans, and drilling or hoeing was most unusual. On the Heath paring and burning[1] was practised in the open fields and there was a tendency to grow too many successive grain crops. The quality of livestock was low. Young described the cattle as 'wretched', for farmers were 'without the least thought of selection'. The result was that when sold they brought half the price of cattle reared on enclosed land. But not only was there little sign of improvement on the South Lincolnshire open fields, but the relaxation of regulations led to a decline in farming standards. This was most noticeable where the fallow had been abandoned and grain crops taken without interruption. In a number of parishes where this had been done the soil was becoming exhausted by the 1790's and crop yields were falling. The poorness of open-field farming struck one visiting Scots farmer most forcibly. He wrote,

[1] The turf was broken by a breast plough, the sods burnt in piles and the ashes scattered over the field.

in 1805, of a village near Grantham: '...the condition...was sufficient to provoke the most patient agriculturalist in the three kingdoms, to have such fine land laying in ridges and bulks alternately, one unproductive, the other part waste, was such an instance of high treason against the good of the community that the possessors deserve to be indicted as public nuisances'.[1]

In Holland the open fields did not survive as a barrier to good farming, but the persistence of common fen, and particularly the practice of intercommoning prevented either effective drainage or improved farming methods. Whilst the interior fen was undrained and unenclosed summer grazing was almost the only possible use of the land. Regular winter floods prevented winter grazing and the absence of private property rights made arable cultivation difficult. As a consequence,

...the higher and more profitable part [the Townland] is made subservient to the unprofitable part of the Estates...the lower parts...when dry give a luxuriant herbage...they are almost universally adapted to the summering of immense quantities of horses and young cattle for the winter support of which they furnish nothing. Hence a sort of necessity arises for the tenants to mow such of their lands as are most out of reach of winter floods...and as a consequence these lands are much weakened by constant mowing....

Intercommoning, of course, made selective breeding virtually impossible; and after the cattle had been turned out on the common fen in summer they received little, if any, attention. In Deeping Fen four-fifths of the cattle grazed there were often lost each summer, and in East, West and Wildmore Fens conditions were equally bad.[2]

It is perhaps not surprising to find that farming on the open fields and common land was so backward; such was true of most of England at this time. But when the enclosed areas are examined—and they were the greater part of South Lincolnshire in 1790—they are found to be little better. However, an important qualification to this must be made. In most regions *some* excellent farmers were to be found, who had adopted most of the new methods. Arthur Young's

[1] Stone, *General View*, pp. 27–9, 35–6, 53; Young, *General View*, pp. 89, 303; L.A.O. Cragg 1/1; *Annals*, XLIII (1805), 428.
[2] S.G.S./S.R./5/37; Stone, *General View*, pp. 19, 22.

Farming Methods and Productivity

work on Lincolnshire is full of descriptions of the most advanced techniques, and he was greatly impressed by Lincolnshire farmers. 'I have not seen a set more liberal in any part of the kingdom', he wrote, 'industrious, active, enlightened, free from all foolish and expensive show.... I met with many who had mounted their nags, and quitted their homes purposely to examine other parts of the kingdom; had done it with enlargened views, and to the benefit of their own cultivation.'[1] Unfortunately this was far from true of the *majority* of South Lincolnshire farmers. The general level—as distinct from the best—remained little in advance of the middle of the century.

This is not to say that there had been no general improvement. Enclosure had allowed some advances to be made. For example the turnip, which had been hardly grown at all in 1770, was widely cultivated by 1801 (see Fig. 11, p. 74), particularly on the Heath. Artificial grasses were also becoming more common. A Heath farmer writing in 1805 knew of 'no enclosure having taken place, but where clover and grasses have been more or less introduced, and where fallows have been made to occur much less frequently than before'. Nevertheless, 'seeds' were not generally grown, even on the Heath, until the 1820's. But, as Thomas Stone pointed out, 'only a cursorary observer would think the introduction of turnips and seeds sufficient improvement'. For the turnip had to be properly cultivated if its full benefit was to be felt. In South Lincolnshire turnips were hardly ever drilled or hoed; nor were sheep normally folded on the crop. Thus much of their value was lost, for without folding the land was not manured, and without proper cultivation it no longer served as a cleaning crop. Where turnips were not grown beans were the most common fodder crop; but their management was no better than that of turnips. The bean fields of South Lincolnshire were 'broadcast, never hoed and full of weeds'.[2]

So even where new rotations were introduced much of their benefit was lost, and improved rotations were in themselves rare. In the

---
[1] Young, *General View*, pp. 39–40.
[2] Young, *General View*, pp. 115, 122, 126, 130, 132, 138, 224, 240, 258, 374, 378, 388; T. Stone, *A Review of the Corrected Agricultural Survey of Lincolnshire by Arthur Young* (London, 1800), pp. 76, 90, 203, 221–3; idem, *General View*, pp. 39, 43; *Annals*, XLIII (1805), 428; Creasey, p. 367.

clay areas the old open-field rotation of (1) fallow, (2) wheat and (3) beans was retained even after enclosure. In the Interior Fens both Young and Stone agreed that there was no regular system of management. The land was pared and burned and then a crop of coleseed (or rape) was followed by a succession of oats crops until the land was exhausted. The better farmers then laid their land down to temporary grass, but the majority simply pared and burned again. On the Townland farms, where the land was mainly enclosed, the general practice as late as 1815 was to grow two white grain crops followed by a bare fallow, whilst on the Marsh even in 1820 some farmers were taking up to ten successive grain crops before fallowing. The Heath provided some interesting internal contrasts, for it was here that progress was most rapid among the more enterprising farmers. On the other hand the general level of farming was deplorable. Paring and burning was used, not simply as a means of bringing land into cultivation for the first time, but as a regular practice. It was followed by a year under turnips and then a succession of grain crops. Not surprisingly parts of the Heath suffered from soil exhaustion, and in places were being returned to the rabbit warren and gorse they had been under before enclosure. 'At least nine tenths of the agriculture of this large tract of country', wrote Thomas Stone of the Heath and Wolds, 'is barbarous in the extreme.'[1]

If the new rotations had made so little impact it is not surprising to find that other improvements had made little progress. The only generally used manure was farmyard dung. The better farmers prepared this by feeding their cattle in stalls on purchased oil-cake. Indeed so little was the significance of fertilizers appreciated that some fen farmers sold their farmyard dung. Even those who appreciated its value only applied it to the field near the farm, and the outlying fields went unattended. Of artificial fertilizers only lime and crushed bones were used to any extent. But most of the acid grasslands of the clay regions received far too little lime. On the arable clays, the manures were easily washed away after rain except in the rare cases where the land had been underdrained.

[1] Young, *General View*, pp. 114, 117; Stone, *General View*, pp. 15, 32, 38, 46; Marrat, I, 89; P. Thompson, *Collections*, p. 373, also references in previous footnote.

Farming Methods and Productivity

Crushed bones were first used in South Lincolnshire on the Heath, the idea having been brought from Yorkshire in the 1790's. But this was a costly process and it was not until after the wars that farmers began generally to use this fertilizer. On the Heath and in the fens underdrainage was not thought to be necessary; in the clay regions only a few farmers had replaced the surface drainage of ridge and furrow with any form of underdrainage.[1]

Only in breeding, in fact, had there been any general improvement in farming practices in South Lincolnshire by the beginning of the nineteenth century. The old Lincolnshire Longwool was 'an animal that had been formed with few other ideas than the production of wool'. But when Arthur Young visited Lincolnshire, farmers were discussing the relative merits of the Longwool and the New Leicester. Most farmers were producing a cross-breed of the two which gave both mutton and wool, and matured more quickly. By 1799 the cross-breed had replaced the Longwool in Kesteven, but the Longwool remained throughout much of the fenland. By 1820, however, the old Longwool had gone for good even there. There were fewer improvements to cattle. It is true that a number of landlords were experimenting; they brought in animals from Ireland, Scotland and Yorkshire, from Northumberland and even the Channel Islands, but this had little effect on the general standard of cattle breeding. It was not, in fact, until the nineteenth century that the Lincoln Red was bred and became the standard shorthorn of the area.[2]

It is clear then that the general standard of farming in South Lincolnshire was still backward in the 1790's and 1800's. But it must be emphasized that there were some farmers in most regions who had adopted much of the New Husbandry. Consequently it was possible to find farms whose methods were hardly different from those of the Middle Ages next to farms whose techniques would not have disgraced those of the 1860's. This was particularly true of the Heath, but even here the idea that crops and livestock

---

[1] Creasey, p. 365; *Annals*, XXII (1794), 530; Stone, *General View*, pp. 16–17, 29, 42; L.A.O. Jarvis 8, 2 Cragg 1; A. Young, *A Farmer's Tour through the East of England* (London, 1771), I, 469; idem, *General View*, p. 367; *Communications to the Board of Agriculture*, IV (London, 1805), 214–15.

[2] Young, *General View*, pp. 303, 364, 371; Marrat, I, 94, 95; S.G.S./B.S./15.

could be properly integrated had made little progress. Some landlords were pressing it on their tenants, and improvers from other countries were canvassing the idea. Nathaniel Kent acted as an adviser for a Holbeach landlord whose estate still remained primarily under grass. He wrote, 'I am clear that the plough alone can pay the rent I set upon it...and all future improvement upon it must arise from getting the land into better heart by the turnip system of husbandry which will carry double the stock that it can carry in its present condition, and which gives much improvement to land'. But even at the end of the wars most Lincolnshire farmers remained either graziers or followed a primitive arable farming system. The carefully integrated mixed farming of the Norfolk system—which many still consider technically one of the finest systems of farming ever devised—remained only a half formulated idea to most South Lincolnshire farmers.[1]

If the literary evidence is an accurate indication of the standard of farming in South Lincolnshire during the war period, then there can have been little increase in farm productivity in the late eighteenth century. There is no doubt that total output in the area did increase but this was mainly a result of expanding the agricultural area (see below, pp. 70-1). No authoritative statement can be made on changes in output per man-hour, for there is little reliable evidence on the size of the labour force in this period. But as there was no significant introduction of labour-saving machinery, and as the New Husbandry required more labour than the old order, it seems unlikely that there was any substantial increase in labour productivity. Most of the new techniques increased output per acre and the new rotations, the folding of sheep, the use of artificial fertilizers and better farmyard manures all aimed at conserving or increasing crop yields. But as it has been shown that these methods had not been generally adopted in the area, it seems reasonable to suppose that crop yields showed little increase in the late eighteenth century. Fortunately this can be confirmed, for there are unusually reliable statistics on crop yields in South Lincolnshire at this time. In 1795 the fear of invasion prompted the government to inquire into the farming of the maritime counties. Returns made by the

[1] L.A.O. T.Y.R. 4/1/103.

Farming Methods and Productivity

Kesteven Justices of the Peace have survived which give the acreage and output for wheat, barley and oats for the years 1792–5. These figures are shown below in Table 3 A, whilst in Table 3 B some less reliable figures on crop yields in the area are shown. Table 3 A suggests that a figure of 16 bushels was an average yield for wheat in Kesteven. There can have been little increase since the middle of the eighteenth century, for whilst there are no estimates at that time for South Lincolnshire, estimates of the average yield of wheat for England were 16 bushels or more.[1] This then seems to confirm the literary evidence; only locally were there increases in crop yields between 1750 and 1795.

TABLE 3. *Yield in bushels per acre, for wheat, barley and oats*

Area	Date	Wheat	Barley	Oats
		A		
Ness	1796–1800	13	24	41
Grantham Soke	1792–5	17	22	28
All Kesteven	1792–5	16	28	33
		B		
The Fenland	1800	48	—	60–112
Kirkby-la-Thorpe	1801	20	24	40
Swarby	1801	20	28	48
Kirkby Underwood	1801	16	—	—
Canwick	1802	24	35	40

Sources: P.R.O./H.O./42/53/95 and H.O./42/37. L.A.O. Brace/14/20/13, P.R.O./H.O./42/53/105 and H.O./67/(1801).

Because both the statistical literary evidence has been drawn primarily from the period 1790–1800, it could be argued that crop yields might have risen in the later part of the wars. There are, however, good reasons to suppose that this was not so. First of all the literary evidence dating from the later parts of the wars shows that there had been no general advance in the standard of farming. Secondly, it seems that continuing high grain prices persuaded many farmers to relax the rules of good husbandry for short-term gains. The result was that on the Heath and in the fenland soil exhaustion occurred and crop yields declined. William Marrat wrote in 1814 'the turfy land of the fens and the lighter silts of the marshes,

[1] Bennet, p. 12.

have much of them by paring and burning and repeatedly [growing] exhausting crops, almost been reduced to a caput mortuum, a state in which even rest from tillage will not improve them'. He estimated that in parts of the Interior Fens there had been a fall in crop yields of about 40 per cent between the first cultivation of the reclaimed peat and the end of the Napoleonic Wars. Lastly, only the poorer land remained to be reclaimed after 1800, and this would inevitably have brought the average yield for the whole area down.[1]

The statistics on wheat yields so far discussed have referred to Kesteven as a whole. But there were significant differences in yields *within* that area, and also between Kesteven and Holland. First of all there were the differences between the areas where farming remained backward and the rarer instances of good farming. Thus in the Hundred of Ness, where there were excellent soils but poor farming, wheat averaged 13 bushels an acre. In Canwick parish on the other hand, which consisted largely of poor Heath soils, but was well farmed, wheat averaged 24 bushels an acre (Table 3A, B, p. 59). More significant, however, were the regional differences in crop yields. Fig. 9 is an attempt to show this. It is based on the examples of crop yields on different types of soil given in the contemporary literature. It shows that there were marked differences in the wheat yields found on the Heath, the clays and in the fen. The highest yields were found in the fenland. This is not surprising as the peat soils had a much higher nitrogen content than any of the other soils of South Lincolnshire. Furthermore, much of the land had only recently been ploughed after being under grass for a long time. Indeed the yields obtained in the fenland probably exceeded those to be found in any other part of England. Outside the fenland the clay soils gave the highest yields. In spite of the difficulties of working the clays, they had a high humus content which was easily maintained and so gave good yields of wheat and beans. On the limestone and gravel soils, however, the humus content was low and without heavy manuring these soils gave poor yields. This pattern of crop yields is then much as might be expected, for as long as farming methods remained backward, crop yields were primarily a function of 'inherent fertility'.

[1] Stone, *Review*, p. 221; Marrat, II, 89–91.

Farming Methods and Productivity

The statistical evidence on yields bears out the literary evidence on the standards of farm management; by the end of the Napoleonic Wars farming in South Lincolnshire was still remarkably backward. This is particularly surprising when it is remembered that not only

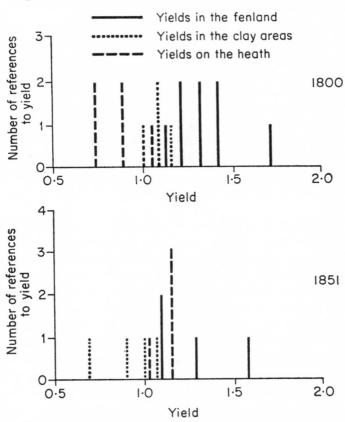

Fig. 9. Yields of wheat, 1800 and 1851. The average yield of wheat on clays is represented as one for each year. Other yields are shown proportionally. In the vertical axis, the number of references of a given yield are plotted.

had most of the area been enclosed by 1800 but that there were certainly some farmers in the area who were improving their farming practices. It is true that one might expect a lag between enclosure and the introduction of new methods; but in South Lincolnshire the lag seems to have been of considerable length. There must

The Agricultural Revolution in South Lincolnshire

also be some explanation of why the majority of farmers had failed to adopt methods which were known and practised by a minority.

A number of factors can be suggested which help explain these circumstances, some affecting the whole of South Lincolnshire, others specific to particular regions.

To begin with the maximum agricultural area was not attained until 1825. There was still land used for little more than rough grazing which could be reclaimed. From 1760 until 1814 grain prices were generally favourable so that a farmer who wished to increase his income had two major possibilities. He could adopt the new methods and increase his yields per acre; or he could increase his output simply by ploughing his poorer grazing land. Normally the latter was the simplest and the cheapest, for it required no experimenting with unfamiliar techniques nor did it need much extra labour. Thus, as long as there remained the possibility of adding to the agricultural area, there was little incentive to introduce the new methods.

Thorough cultivation, underdrainage, the use of artificial fertilizers and farmyard manure were fairly expensive. If a tenant was to undertake the type of improvement which was semi-permanent in its effect he needed a guarantee that he, rather than a succeeding tenant, would derive the benefit. In Norfolk tenants were encouraged to improve by being given long leases; in Lincolnshire, however, the great majority of tenants were on an annual tenancy and could be evicted with only 6 months notice. They received no compensation for unexhausted improvements on leaving the farm. Consequently tenants had no security for improvements which they undertook, and there is no doubt that this was an important factor in explaining the backwardness of South Lincolnshire, particularly in comparison with the neighbouring county of Norfolk.[1]

In any case the farmers in the area seem to have been—with a few exceptions—singularly unenterprising. A Bourne landlord described his neighbours and tenants as 'an unenlightened peasantry' and it seems that his description, if uncharitable, was more fitting than Young's enthusiastic eulogies (see above, p. 55). For example, the tenants on Colonel Jarvis's estate at Doddington, as described in

[1] Stone, *General View*, pp. 39–40, 96; *idem, Review*, p. 60; Young, *General View*, p. 59.

Farming Methods and Productivity

1811, present a dismal picture: '...as to the present occupiers continuing the land, no one can ever think of it who has seen their mode of management. There is some of the finest turnip land in the county, but the portion that is sown with turnips has never been hoed nor is the land gripped to take off the surface water—for want of this the sheep at this time, upon the turnips, are up to their knees in puddle.'[1]

If the tenants lacked security and enterprise—and the two were clearly interrelated—landlords were generally little better. A major explanation of the differences between Lincolnshire and Norfolk farming was the failure of Lincolnshire landlords to support their tenants' improvements after enclosure. In Norfolk landlords not only paid the cost of enclosure but bore part of the expenditure on improvements such as marling and draining. In Lincolnshire this was not so and consequently the mere fact of enclosure did not necessarily lead to improvement. But before Lincolnshire landlords are judged too harshly it must be remembered that they had contributed heavily to enclosure, fen drainage and the construction of canals, and they may well have lacked the resources to finance further improvements. During the war period they increased rents far less proportionally than the increase in prices, and the tenants retained a higher proportion of gross farm income. It was estimated that in England as a whole the landlord received in 1790 a third of gross farm income, but only a fifth by 1815. In Lincolnshire landlords may have experienced an even greater proportionate loss.[2]

The factors considered so far applied throughout the whole of South Lincolnshire; but there were a number of reasons why certain regions should have lagged behind the area as a whole. Most of the innovations in agriculture were in arable farming and only indirectly affected grazing. Yet much of South Lincolnshire remained grazing land down to the end of the Napoleonic Wars. As late as the 1820's it was noticed that there was a considerable difference in the standard of farming in the arable and grazing districts. Parts of South Lincolnshire had been grazing regions for a long time, and even where farmers were turning to arable farming, they had little

---

[1] *Annals*, XXXIX (1803), 558; L.A.O. Jarvis 8.
[2] Stone, *General View*, pp. 39–40; B.P.P. III (1813–14), 198.

experience in its methods. This was noticeably so in the fenland. In 1781 an agent of Lord Brownlow's wrote that his tenants, 'having employed themselves in grazing [they] are more competent for the occupation of grass than arable'. Forty years later a Skirbeck farmer described the deplorable standard of arable farming in Skirbeck Hundred and attributed it to the graziers' lack of experience.[1]

There were significant regional differences in soil type as well as in land use, and this too affected the regional rates of progress (see above, chapter 1). The Norfolk system was based on the growth of turnips, seeds and the folding of sheep. On clay soils turnips were rarely successful and on undrained clays sheep suffered from foot rot. Consequently there was little progress made during the wars in the clay regions. But on the limestone and marlstone soils of the Heath and on the gravel soils of north-west Kesteven the turnip could be grown and sheep folded. Whilst it is true that these regions were still generally badly farmed, it was certainly here that most progress had been made; as Young succinctly put it, 'They are awake and moving on turnip land; but on bean soils, are still fast asleep'. The fen soils, with the exception of the lighter silts of the Marsh, were not generally suited to turnips. More important, however, was that fen farmers believed their soils to be of unlimited fertility. 'The quality of this kind of land is almost inexhaustible', wrote one of them. Consequently there was less incentive to introduce new rotations, whose essential purpose was to conserve soil fertility. It took a generation of bad farming and declining yields to disabuse fen farmers of this view.[2]

There were other factors which may have affected the differential rate of regional improvement, although their importance is difficult to assess. Many agricultural writers in the eighteenth century believed that it was easier to adopt the new methods on the large farms of the Wolds than on small farms.[3] If there was a significant difference between the rate of improvement on large and small farms, then this would further account for the differing progress of west Kesteven, the Heath and the fen, for the Heath had many large

---

[1] L.A.O. 2B.N.L.; P. Thompson, *Collections*, p. 371; Creasey, p. 366.
[2] Young, *General View*, p. 84; P.R.O./H.O./42/53/8.
[3] Young, *General View*, p. 38.

Farming Methods and Productivity

farms whilst in west Kesteven and the fenland farms were predominantly small (see below, pp. 91–4). A further significant difference was between the occupier owner and the tenant farmer. The occupier owner had to bear the cost of all the improvements on his farm, whereas the landlord bore the cost of permanent improvements on a tenanted farm, leaving the tenant free to utilize his capital on short-term improvements. As has been noticed earlier, neither tenant nor landlord seems to have been prepared to invest substantially in improvements other than enclosure, fen drainage or the reclamation of marginal land; but in the cases where there was a happier relationship between tenant and landlord the Heath again had the advantage, for the great majority of occupiers there were tenants; in the fens and west Kesteven, however, there was a substantial minority of occupier owners.

There seem quite adequate reasons then to explain why there had been so little progress made in South Lincolnshire farming at the end of the Napoleonic Wars; and further to explain why the Heath had shown most progress in the area as a whole.

CHAPTER IV

THE PATTERN OF LAND USE

Today more than four-fifths of the agricultural area of South Lincolnshire is under arable, and less than a twentieth of the total area agriculturally unproductive.[1] But at the beginning of the nineteenth century the county was celebrated for its pastures—'the glory of Lincolnshire'[2]—and notorious for its barren heaths and undrained fens. This latter ill-repute was undeserved, for great strides had been made by enclosure and drainage in the last quarter of the eighteenth century. Nevertheless, there was still a great deal of land which was at the best only rough grazing. These 'wastes', as they were somewhat inaccurately called by contemporary writers, occurred principally in two regions; on the Heath and in the Interior Fenland.

In the 1770's the Heath was 'a bleak, poor and unsheltered open field country, under a miserable system of cropping; and a very considerable tract was kept in warren'. The rising grain prices of the war period made the enclosure of even these poor soils profitable. By 1815 the open fields and their commons had disappeared, but waste still remained. A traveller from Sleaford to Lincoln in 1801 described the region as one of 'barren heaths and rabbit warrens', whilst 15 years later a lady making the same journey thought the Heath a 'living picture or a moving plain of a busy republic of rabbits'. In much of the Heath waste was ploughed up immediately after enclosure, but often farmed so badly that soil exhaustion soon set in. By 1800 parts of the Heath had been returned once again to gorse or rabbit warren. In the 1790's there was still wasteland in north-west Kesteven and for the same reason as on the Heath; the gravel soils were so poor that only very high grain prices justified farming them. A good deal of land here remained covered with gorse as late as the 1840's; and during the wars

---

[1] *Agricultural Statistics, 1957/1958* (H.M.S.O., London, 1960), pp. 28–30.
[2] Young, *General View*, p. 174.

The Pattern of Land Use

at least several landlords found it more profitable to have part of their estates in plantation rather than poor grazing.[1]

The second area of extensive waste was the Interior Fenland. Until 1801 East, West and Wildmore Fens, Deeping Fen and smaller parts of the Witham and South Holland Fens were so badly drained that they could only be used for summer grazing; even then large numbers of the livestock were lost every year. By 1801 the whole of the Interior Fens had been enclosed, but their drainage was far from uniformly successful. The result was that not only did land use vary from parish to parish, but also from year to year. Immediately following enclosure and drainage the fens were invariably ploughed. But the continued danger of flooding meant that there was a considerable risk attached to arable farming. Some farmers compromised by sowing only a spring crop of oats or coleseed; but the success of this form of catch cropping depended on getting the spring flood waters off the land early enough, and the autumn floods coming late enough, for the harvest to be gathered in. These then were the 'half-yearly lands' which John Rennie wrote of in 1807 (see above, p. 27); and this was the condition of the greater part of the Interior Fen throughout the Napoleonic Wars. Some farmers, however, were constantly tempted by the high prices, and ploughed their fertile fen for year after year. A tenant of the Earl of Bristol, for example, only returned his fen lot to grass after he had lost his harvest to floods for six consecutive years. Even in the last few years of the war period only East, West and Wildmore Fens had been successfully drained and converted *permanently* to arable.[2]

Permanent grassland was to be found throughout South Lincolnshire, but there were three regions where it was the dominant form of land use: the Townlands, south-east Kesteven and west Kesteven (Fig. 8, p. 51).[3]

---

[1] Creasey, pp. 365–6; *Farmer's Magazine*, 1 (1800), 394; Miss S. Hatfield, *The Terra Incognita of Lincolnshire* (London, 1816); Young, *General View*, pp. 234, 249–50; L.A.O. Jarvis 8, Cragg 1/1.

[2] L.A.O. 2 Cragg 1/5; *J.R.A.S.E.* XII (1851), 295; P.R.O./H.O./67; S.G.S./B.S./5/37.

[3] The map and the following description are based on: John Cragg, *A Topographical Dictionary of Lincolnshire* (a number of manuscript versions of this unpublished work are in L.A.O. Cragg 1/1); Marrat; Thompson, *Collections*; Young, *General View*, pp. 174–7, 186, 189, 191, 208, 234, 388, 432; Stone, *General View*, pp. 32, 38, 76, 221–3; *Communications to the Board of Agriculture*, IV (1805), 50.

The Agricultural Revolution in South Lincolnshire

Between Wrangle and Tydd St Mary well over two-thirds of the Townland was under grass, and there were in addition good pastures on adjacent parts of the Marsh and the Interior Fens. This region carried the highest average rent in Lincolnshire—some pastures near Boston let at nearly £3 an acre—and were among the best grazing lands in England. William Marrat, writing in 1814, was quite clear as to the proper use of this land.

> Nothing but a sudden and pressing demand for an increased production of corn could ever occasion any additional quantity of this land being brought into cultivation. For taking the whole kingdom as one large farm (and for the real and permanent interest of the nation it ought to be cultivated as such) the natural apportionment of this whole level is to pasturage and feeding.[1]

There were two reasons for the high quality of these pastures. First the silt soils naturally give a good grass. A grassland survey of eastern England in 1939–40[2] showed that all the pastures of Holland fall into the first three grades of grass that the survey recorded, a standard attained by no other county. Marrat and other contemporary writers noticed too that whilst in the summer months the livestock of the Holland parishes were grazed on the Marsh, the Townland and the Interior Fens, they were wintered solely on the Townland, which for generations had received the benefit of their manure.

Nowhere else in South Lincolnshire were there pastures to match those of the fenland. Nevertheless, the grasslands of the extensive grazing district in south-east Kesteven were good by any standards and certainly envied by neighbouring farmers. The clays near Folkingham, one such farmer wrote, 'are clothed in a superfine coat of deep green... had I a farm there I would be very wary of sticking my plough into land of this description'. Possibly three-quarters of the agricultural land in this district was under pasture, and in some parishes no more than a seventh of the land was ploughed.

The third grazing region was in west Kesteven, stretching from the foot of the limestone and marlstone scarps towards the Trent. There was probably a lower proportion of land under grass than in

---

[1] Marrat, II, 88; Young, *General View*, p. 189.
[2] Stamp, parts 76 and 77, pp. 521–5.

The Pattern of Land Use

the Townlands or south-east Kesteven, and the pastures were certainly poorer except in the south, where they merged into the celebrated grazing lands of the Vale of Belvoir.

There were a number of reasons[1] why these three regions should have been mainly under grass, but the most important was their soil type. All three districts were on heavy clays (cf. Figs. 2 and 8, pp. 15 and 51) and in the eighteenth century clays were traditionally under wheat or grass. The heavier clays were not only difficult to work, but gave a good grass, noticeably in comparison with the lighter soils adjacent to these three districts. The early date of enclosure suggests these regions had a long tradition of grazing. Indeed the fact that they had been enclosed so early helped explain the continued predominance of grass. Parishes enclosed by Parliamentary act invariably had their tithes commuted to a rent charge, but old enclosed parishes continued to pay tithe in kind. Payment in kind was not only a deterrent to improvement, but acted as a sort of subsidy on pasture, for a farmer paid more in kind on arable than grass. Young and Stone disagreed on the significance of this factor, but there is no doubt that it persuaded some farmers to keep their land in grass. Nathaniel Kent, after valuing a Holbeach estate, bluntly informed the landlord that 'the only aim of your policy at present is to avoid tithes'. But even where tithes were not paid in kind the eighteenth-century landlord put a premium on pasture, which invariably bore a higher rent than arable. Most tenancy agreements forbade the ploughing of permanent grassland and heavy fines were imposed if this was disobeyed. One reason for this concern was the difficulty of restoring pasture satisfactorily once it had been ploughed. Lastly, the influence of accessibility should be noticed. Cattle were easier to market than grain where transport conditions were bad. When John Cragg wrote his Topographical Dictionary in the 1800's, he observed that a number of parishes in south-east Kesteven were mainly under grass for this reason (cf. Figs. 6 and 8, pp. 42 and 51).

At the beginning of the war period there was probably no region in South Lincolnshire where arable was the predominant form of

[1] P.R.O. H.O./42/53/99; Young, *General View*, p. 77; *Farmer's Magazine*, II (1801), 254; L.A.O./T.Y.R./4/1/93, 2 Cragg 1, Cragg 1/1.

land use. Even by 1801 tillage occupied no more than a third of the area.[1] Nevertheless, the high prices of the war period led to a remarkable expansion of the arable acreage. In 1796 the Kesteven Justices of the Peace were asked to make a return of the acreage and output of grain crops in the county. They made estimates for the years 1792-5, and these figures, together with the acreages returned to the Home Office in 1801, are shown in Table 4. The figures show a striking expansion in the arable acreage between 1792 and 1801 and this was probably continued down to the end of the war period. In Holland the greatest increases in arable land came after 1801; there were numerous contemporary witnesses[2] to the great increase in the fenland in this period, and this can also be deduced from Table 5, which shows the export of grains from Boston between 1803 and 1815.

The increases in arable land were, however, not equally distributed. In Kesteven most of the new arable land followed the enclosure of parishes on the Heath when rabbit warrens, waste and rough grazings were ploughed up. Some of the pastures of west and south-east Kesteven *may* have been broken up, but it seems that the institutional restraints (see above, p. 69) prevented the conversion of much grassland. In Holland the only part of the Interior Fens permanently converted to arable was in the north. The drainage of East, West and Wildmore Fens was completed in 1807, and the land at once put under the plough. The impact on grain production is clear enough in Table 5. South of the Witham the Interior Fens were rarely successfully converted to arable; for example, Deeping Fen was completely under water in 1800 but the whole 15,000 acres were sown to oats and seeds in 1801.[3] Much of the new arable land in Holland came in fact on the Marsh. George Maxwell, who appeared before a Select Committee in 1814, was at great pains to make clear that the arable increases in

---

[1] The parish estimates of crops made in 1801 omitted artificial grasses and certainly underestimated other crop acreages; the South Lincolnshire returns covered 77 per cent of the area. The total recorded acreage came to 117,492 acres. Even allowing for the underestimates and omissions the actual acreage was unlikely to have been more than one-third of the total acreage of Kesteven and Holland—728,939 acres.

[2] For example, Marrat, I, 88; P. Thompson, *Collections*, p. 381.

[3] Stone, *Review*, p. 131; L.A.O. 1 Cragg 2/30; P.R.O./H.O./67/Lincs.

The Pattern of Land Use

TABLE 4. *The acreage in Kesteven under grain crops, 1792–1801*

	Wheat	Barley	Oats	Rye	Total
1792	15,846	13,165	16,615	360	45,986
1793	15,595	18,568	17,582	373	52,118
1794	18,614	19,352	21,202	489	59,657
1795	16,126	15,773	23,672	301	56,071
1801	(a)17,018	23,305	26,743	347	65,413
1801	(b)26,500	28,960	33,430	540	89,430

Sources: 1792–5, P.R.O. H.O. 43/37. Returns to the Duke of Portland by the Justices of the Peace of Kesteven, 1796. 1801, P.R.O. H.O. 67/Lincs. The first line for that year (a) gives the actual figures in the returns. The second line (b), includes estimates for the parishes without returns.

TABLE 5. *Number of quarters of grain exported from Boston, 1803–15*

	Wheat	Oats	Barley
1803	5,190	191,048	924
1804	7,676	179,553	456
1805	2,513	201,898	369
1806	3,505	257,864	1,499
1811	32,638	360,699	570
1812	45,238	251,504	140
1813	43,985	239,063	784
1814	15,105	254,916	0
1815	22,275	246,160	183

Source: P. Thompson, *The History and Antiquities of Boston* (Boston, 1856), p. 350.

the fenland had come mainly, not from the Interior Fens, but from the Marshlands which lay seaward of the Townland.[1] These lighter silts still had in the 1790's a great deal of poor sheep grazing land as well as arable and grass; by the end of the wars they were predominantly under arable. There was some ploughing of the fine pastures of the Townland, but as in Kesteven the greater part remained untouched until the end of the wars.[2]

Although it is not possible to describe the distribution of arable land with any precision, crop distributions can be discussed with more assurance. In 1801 clergymen were asked to make an estimate of the acreages under different crops in their parishes. These returns

[1] B.P.P. III (1813–14), 242.   [2] Marrat, I, 88.

were eventually forwarded to the Home Office and form the first reliable account of crop distributions in England and Wales.[1] In South Lincolnshire returns were not made for only a few parishes and the surviving records cover nearly four-fifths of the area. No estimates were made of fallow, permanent or temporary grasses; and in the Lincolnshire figures, peas and beans are often recorded as one figure whilst rape and turnips are in all cases so returned. However, as rape was hardly grown at all outside the fenland and turnips rarely within it, this is not greatly important. It is generally agreed that whilst the 1801 returns certainly underestimated the actual acreages, they give a fair representation of the relative importance of crops within each parish.

TABLE 6. *The 1801 crop returns*

Crop	Total recorded acreage	Crop as a percentage of total recorded acreage*
Oats	38,549	33
Wheat	24,566	21
Barley	24,507	21
Turnips and rape	18,697	16
Peas and beans	9,419	8
Rye	347	—
Potatoes	1,407	1
Total acreage	117,492	100

* The percentages are rounded.
Source: P.R.O./H.O. 67/Lincs.

The aggregate figures for South Lincolnshire are shown in Table 6 and the distributions are in Figs. 10 and 11. The most striking feature of the aggregate figures is the dominance of grain crops, which occupied three-quarters of the recorded acreage. Of the individual crops oats was by far the most important, and wheat and barley shared second place. It is not possible to determine the

[1] The use of the 1801 crop returns has been discussed by H. C. K. Henderson, 'Agriculture in England and Wales in 1801', *Geographical Journal*, CXVIII, no. 3 (1952), 338–45; W. E. Minchinton, 'Agricultural Returns and the Government during the Napoleonic Wars', *Agricultural History Review*, I (1954), 2, 9–43. See also D. B. Grigg, 'The 1801 crop returns for South Lincolnshire', *The East Midland Geographer*, 16 (Nottingham, 1961), 43–8.

The Pattern of Land Use

acreage under rape and turnips separately, but combined they occupied twice the area which beans and peas did.

Oats were widely grown, but were of greatest importance in the fenland where they occupied over half the recorded acreage. They were grown especially in the Interior Fens for the crop was more

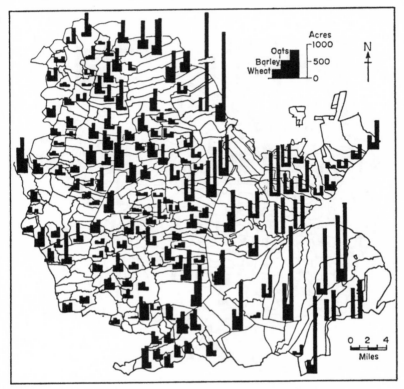

Fig. 10. The distribution of grain crops in 1801.

suited than either wheat or barley to the newly cultivated peat soils. The paring and burning of the turf was invariably followed by a series of oats crops. The crop served the same 'pioneer' function on the Heath and was also grown on the poorer soils of Kesteven, on the sandier limestone soils near Grantham and particularly on the gravel soils of north-west Kesteven. Rye, which occupied a negligible acreage was almost entirely confined to this latter region.

The Agricultural Revolution in South Lincolnshire

Oats were used mainly as a fodder crop, particularly for horses. In 1798 there were over 15,000 horses in South Lincolnshire, where they were replacing oxen on the farms and, of course, were still the main means of general transport. Feeding them was a

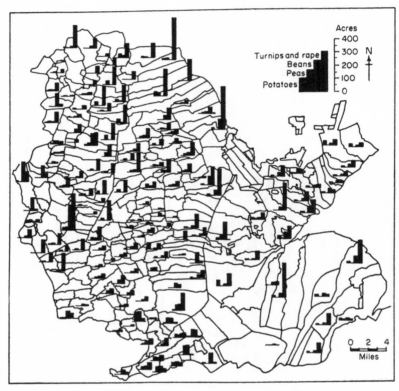

Fig. 11. The distribution of fodder crops in 1801.

major problem, and at least one farmer seems to have begrudged them their food. During the grain shortage of 1801 John Mossop wrote to the Duke of Portland that 'it is to be hoped that the horses and pigs may not be allowed to swallow so large a portion of grain as they have done in more plentiful times...throughout the lowlands of Lincolnshire...some of the very best land in the Kingdom ...the number of horses grazed [is] almost incalculable'. But not

The Pattern of Land Use

all the output was locally consumed, for much of the enormous fenland production was exported, primarily to London.[1]

Barley and wheat, whilst occupying similar acreages had very different distributions. Wheat was grown in nearly every parish, but nowhere was it the dominant crop as oats was in the fenland. It was the major crop in only two regions (Table 13, p. 98), both in areas of clay soils in Kesteven. A third of the total production came from Holland where the crop everywhere took second place to oats. At the beginning of the war period the bulk of the Holland wheat crop was grown on the heavier silts of the Townlands and the Marsh. It was rarely grown in the Interior Fens. This was because when the reclaimed peats and skirty soils were first cultivated, they gave a poor quality wheat crop. But after a series of oats crops, which exhausted the soil, a better quality wheat could be grown.[2] This had happened, for example, in Holland Fen. When first cultivated in the 1770's little wheat was grown, but by the 1790's wheat was overtaking oats as the major product.[3] This process was going on throughout the fenland during the wars, and is reflected in the exports of grains from Boston (Table 5, p. 71). Between 1803 and 1815 the volume of wheat exported rose by over 300 per cent, but exports of oats increased by only 31 per cent. This was not happening in Kesteven. Table 4 shows that it was barley and oats which expanded most rapidly between 1792 and 1801, because most of the new arable was on the limestones, where oats and barley yielded better than wheat.

Barley was much more localized in its distribution than wheat or oats. It was hardly grown in the fenland at all, for even on the lighter silts it gave a very poor quality malt. The best malting barley soils were those of the limestone Heath, and it was here that the crop was most important, occupying a third of the arable. Rather surprisingly, for clays were not generally regarded as good barley soils, it was widely grown in west Kesteven. However, much of the arable land in this largely grazing district was on the patches of

[1] P. Thompson, *History*, p. 109; P.R.O. 42/53/95.
[2] Soils with an excessive nitrogen content give a wheat more straw than grain; however, unless a rotation is followed or fertilizers used, the nitrogen content of any newly ploughed soil soon declines.
[3] *Annals*, XIX (1793), 188.

gravel to be found in most of the parishes. In addition there were a number of maltsters in the area, particularly in Newark. Farmers who sold their barley in Newark brought back lime and applied it to their clay soils. This, according to Arthur Young, gave a malting barley as good as that grown on the limestones. There were, however, risks attached to growing barley on clay soils. In very wet years the quality of malting barley on the undrained clays was so poor that maltsters often refused to accept the crop.[1]

Even allowing for the omission of rotational grasses from the returns, it is clear that green crops occupied a small proportion of the arable land in 1801. Nevertheless, the turnip was already widely grown in the area, considering that it had been hardly grown at all in the 1770's.[2] The crop was, however, confined largely to the lighter soils, which were not only the most suitable for its cultivation, but also most in need of the benefits turnip cultivation brought to poor soils. On the Heath the crop occupied a fifth of the recorded acreage, and it was of equal importance on the gravel soils of north-west Kesteven. Fig. 11 (p. 74) and Table 13 (p. 98) suggest that the crop was also grown on the clays of west and south-east Kesteven; but it is more probable that the turnip was to be found on the small patches of lighter soil which occurred within these regions.[3] In the fenland the turnip was of minor importance, not being grown at all in the Interior Fens and only on the lighter silts of the Marsh and Townland. The main fodder crop in the fens was coleseed or rape, which like the turnip was fed off by sheep; whilst like oats it was particularly suitable for recently reclaimed land.[4] Beans, the normal green crop in the three-field system, remained in South Lincolnshire primarily on the heavy clays of west and south-east Kesteven and on the Townlands. The crop occupied in addition a fifth of the arable land in the villages around Market Deeping. The soil here, a

[1] Young, *General View*, p. 260; *Annals*, XXXI (1798), 203; B.P.P. II (1806), 382.
[2] Young, *General View*, p. 115.
[3] There were river gravels in most of the west Kesteven parishes and outcrops of limestone in every parish in south-east Kesteven. Contemporary literature together with the evidence of the Tithe Redemption maps of the 1840's suggest that the heavier soils in these parishes were mainly under grass, the lighter soils arable (see L.A.O. Cragg 1/1; Young, *General View*, p. 113).
[4] Rape or coleseed could be sown in May and stocked six weeks later (G. E. Fussell, 'The History of Cole', *Nature*, London, CLXXVI, 1955).

The Pattern of Land Use

sandy loam, was eminently suitable for the cultivation of turnips; the reason why beans still predominated was that in 1801 these villages still had open fields.

The 1801 returns made no mention of artificial grasses, a measure of their unimportance in the country as a whole. However, they were certainly grown by some farmers in nearly every part of South Lincolnshire, but not as yet by more than a minority of farmers. They were an essential part of the Norfolk four course, for clover restored the nitrogen content of soils, and consequently they were particularly necessary on the lighter soils of the Heath. It is an index of the relative backwardness of Lincolnshire farming that a year under 'seeds' was not general on the Heath until the 1820's.[1]

At the end of the eighteenth century the potato was probably only grown in any significant quantities in England in the counties of Lancashire and Lincolnshire;[2] and in Lincolnshire the bulk of the crop was grown in the fenland. At the beginning of the war period the crop was utilized primarily as a cattle feed; but rapidly rising wheat prices led to a search for a cheaper substitute for wheaten flour. Rice, barley and oats were all tried in South Lincolnshire, but the potato was regarded as the most acceptable alternative. A Boston correspondent of the Duke of Portland wrote: '...potatoes are much grown and used as an excellent substitute everywhere in the county; and I presume a better substitute is not known nor perhaps never will be'. The Duke's correspondent would doubtless have been surprised at the accuracy of his prediction. When he wrote there were little more than a thousand acres under potatoes in Holland; 70 years later there were over 16,000.[3]

A number of minor crops were grown in the fenland in small quantities but were not recorded in the 1801 returns. Flax and hemp had been of considerable importance in the middle of the eighteenth century but the lifting of the tariff on imports had precipitated a decline in the acreage. Most of this had been grown by small farmers, and when grain prices soared in the war period many

[1] Young, *General View*, pp. 131–4; L.A.O. Brace 14/20/14; Creasey, p. 369; P. Thompson, *Collections*, p. 379.
[2] Henderson, p. 342.
[3] P.R.O./H.O. 42/37/53; *Annals*, XXIV (1795), 120; Young, *General View*, pp. 142, 147; Board of Agriculture, *Parish Abstracts, 1875, Guildford*.

of these farmers were compelled to grow their own wheat for flour—which they had hitherto bought from millers—and consequently reduced their flax and hemp acreages even further. Two other unusual crops were mustard seed, which was grown almost exclusively on newly reclaimed marshland, and woad. Woad attracted more attention from contemporaries than its acreage merited, for it was grown in very few parishes. A number of farmers leased small acreages to woad companies from Yorkshire. Extraordinary rents were paid, for the fen farmers not only claimed that the crop quickly exhausted the soil but also that they found it difficult to return the land to good pasture. Thus John Parkinson leased land to the Yorkshire Woad Company at £5 an acre in 1800; a price well above that paid for even the best pastures in the district. The most celebrated woad grower was John Cartwright of Brothertoft, who devoted 200 of his 1100 acres to woad. It was all sold in the West Riding textile district.[1]

'The management of arable land in Lincolnshire', wrote Arthur Young, 'has never been celebrated.'[2] And certainly throughout the war period Lincolnshire farmers were still more concerned with pastures and livestock than the niceties of arable farming. Sheep, cattle and pigs were found everywhere, as they are today. More unusual were the large number of oxen and horses which were to be found on the farms of the area (see Table 7). Although oxen were being replaced by horses as work animals, there were still nearly 12,000 at the end of the eighteenth century.[3] Horses in fact only just outnumbered oxen, although there were in addition nearly 3000 kept for purposes other than farming. Sheep were by far the most numerous animals. The total number was less than in Victorian times, but by then the agricultural area had substantially increased. Thus the sheep density could not have been *greatly* less than it was in the 1870's. Cattle were, however, rather less numerous than later in the century; the cattle-sheep ratio in 1798 was approximately 1/8, in 1875 it was 1/6½. The cattle were

[1] *Communications to the Board of Agriculture*, IV (1805), 187; Marrat, 1; P.R.O./H.O./67/Lincs.; Young, *General View*, pp. 147, 157, 161, 167, 198; L.A.O. 2 B.N.L. 15.
[2] Young, *General View*, p. 92.
[3] *Ibid.* pp. 377–80.

The Pattern of Land Use

kept primarily for beef, dairying being unknown except for farm consumption.[1]

Whilst sheep and cattle were to be found in every region in South Lincolnshire, the greatest densities were in the three grazing districts of south-east Kesteven, the Townlands, and to a lesser extent west Kesteven.[2] This is not surprising as few farmers at this time had learnt that the growth of fodder crops, the use of artificial feeds and the practice of stall feeding could maintain a higher stocking ratio than an equivalent area under grass. Sheep densities were particularly high in south-east Kesteven and the Townlands. Table 8 (p. 80) shows the number of sheep per hundred acres of agricultural land for the few parishes for which statistics are available. The contrast between the densities of the two grazing districts and the limestone Heath is clear. There are unfortunately no similar figures for west Kesteven, but they were almost certainly lower. Not only was there more arable here and the pasture inferior, but neither Stone nor Young regarded this region as a major grazing district.

TABLE 7. *Livestock in South Lincolnshire, 1798*[3]

	Oxen	Cows	Young cattle	All cattle	Sheep and goats	Draught horses	Pigs
Kesteven	8,085	10,082	23,819	41,986	306,922	8,907	14,027
Holland	3,638	2,679	7,038	13,355	121,079	3,443	3,795
South Lincolnshire	11,723	12,761	30,857	55,341	428,001	12,350	17,822

The Townlands were pre-eminently fattening pastures for sheep and cattle. Some sheep were bred locally by the smaller farmers, but more usually they were reared on the poorer pastures of the

[1] *Board of Agriculture Returns, 1875, Parish Abstracts, Guildford.*
[2] The description of the livestock economy is based on: P. Thompson, *Collections*, pp. 376–7; Stone, *General View*, pp. 56–7; Creasey, p. 366; Young, *General View*, pp. 176–7, 186–91, 300–3, 337, 341, 364; G. Collins, 'Cattle of the County', *Lincolnshire Magazine*, VII, no. i (1933); L.A.O. Cragg 2/32/1, 2 B.N.L.; Thirsk, p. 302.
[3] The fear of invasion led to an inventory of dead and livestock being taken in 1798 for every parish in the coastal counties of eastern and southern England. In South Lincolnshire returns survive for only one parish, but aggregate figures for Military Subdivisions—groups of wapentakes—allow the totals for Kesteven and Holland to be calculated (*L.A.O. K.C.C. Deposit, Map and Schedule of Livestock and Deadstock in Lincolnshire, 1798*).

Heath and the Wolds. They were then sold by these farmers as shearlings at the spring fairs, and bought by graziers from the Townlands who fattened them, sold their wool and finally sent them off for slaughter. Wool was just beginning to lose its premium, for the new cross-breeds gave good mutton as well as wool. In south-east Kesteven rather more local breeding was done; on the other hand graziers in this region bought their shearlings either from the Heath or from farther afield in Northamptonshire and Leicester. The West Riding, however, was the main market as it was for the Townland graziers.

TABLE 8. *Sheep densities in selected parishes, 1790–1815*
(per hundred acres of agricultural land)

Threekingham	212	(South-east Kesteven)
Folkingham	250	(South-east Kesteven)
Wigtoft	210	(Townland)
Canwick	86	(Heath)
Little Bytham	133	(Heath)

Sources: L.A.O. 2 Cragg 1/6, Brace 14/20/14; W. Cragg, *A History of Threekingham with Stow, in Lincolnshire* (Sleaford, 1912).

Cattle were mainly fattened for beef, and the pastures of the Townland, south-east Kesteven and west Kesteven were admirably suited to this purpose. The density was necessarily less than the sheep density in these regions, and a great deal below what it was when the New Husbandry replaced the old style grazier later in the century. Cattle were also brought from greater distances to be fattened, and as well as local cattle, Irish, Scots and Yorkshire beasts were found here. Similarly, markets were more diverse. Although most beasts were sent to Smithfield, Sheffield and the textile cities of the West Riding bought South Lincolnshire cattle. Some graziers only partly fattened their beasts and sent them on to Norfolk where they were finished before going on to Smithfield.

Sheep and cattle were not, of course, confined to the three grazing districts, although their livestock densities were higher than in the other regions. The Heath had long provided rough grazing for large numbers of sheep, and remained an important rearing region. The

The Pattern of Land Use

pastures were too poor for fattening and farmers were only just realizing that densities could be increased by the adoption of new methods. Until enclosure the Interior Fens were hardly used for any other purpose than summer grazing; but these animals were invariably wintered on the Townlands. After enclosure many Townland farmers retained fen lots which they used for grazing. West Kesteven seems to have kept an unhappy compromise between specialized grazing and indifferent arable farming. Livestock densities were certainly lower than in south-east Kesteven or the Townlands. Yet these pastures still served for fattening some Heath livestock as well as those locally bred.

CHAPTER V

LANDLORDS, FARMERS AND FARMS

The role of the landlord in eighteenth-century agriculture has been frequently discussed and perhaps overemphasized. A landlord could promote improvement amongst his tenants either by example or direction. The landlords of South Lincolnshire seem rarely to have set a good example to their tenants and their home farms were often as badly managed as the farms of their humblest tenants. It is true there were distinguished exceptions, but in general landlords did not give a lead, except in livestock breeding. Nor did they use their opportunities to *direct* tenants towards better methods. A typical tenancy agreement allowed the landlord to specify the rotation to be followed, prevented the tenant from carting hay, straw or manure off the farm, and forbade him to plough ancient pasture. Yet many landlords failed to draw up such agreements or turned a blind eye to infringements. As a whole it can be said that, with the exception of their part in enclosure and drainage, landlords did little to advance the general standard of farming.[1]

Why this should have been is far from clear. However, two possible explanations can be explored. One is that there were few great landowners in the area, and consequently no one with the interest to spread new ideas. The second is that the large landlords lived outside the county and showed little interest in the farming of their land.

The first comprehensive survey of English landownership was not made until 1873, when the 'new Domesday book' was compiled.[2] During the war period the only reliable sources of information are the Land Tax returns, a few estate surveys, and the Topographical Dictionary of John Cragg, which lists the numbers of proprietors in each parish and names the major landowners.[3]

[1] L.A.O. Spital 111/17.
[2] *Return of Owners of Land, England and Wales (Exclusive of the Metropolis), 1873* (H.M.S.O. London, 1875), 1.
[3] L.A.O. K.C.C. Deposit Land Tax, Returns for 1798 and 1808, Jarvis 8, 2 Cragg 1/6 and 8; 5 Anc 4; Holywell 64/3; Cragg 1/1.

Landlords, Farmers and Farms

Imprecise as these data are, they make it clear that there *were* a number of great landowners in South Lincolnshire and that the majority of them did reside there.

The greatest landowner was the Duke of Ancaster, whose home was at Grimsthorpe on an estate of over 20,000 acres. A number of other landlords held at least 10,000 acres: Sir John Thorold, Sir William Welby, Lord Brownlow, Sir Montagu Cholmley, Edmund Turnor, Sir William Manners, Sir Gilbert Heathcote, Sir Thomas Whichcote and Charles Chaplin. Of these, the first five all lived in houses set in the attractive country just north and south of Grantham. Manners, Heathcote and Whichcote all, like the Duke of Ancaster, lived in south-east Kesteven. Only Chaplin, whose home was at Blankney, lived in the northern half of Kesteven. In addition to these considerable owners there were many middling landlords: for example, Earl Fortescue, Lord Boston, Phillip Blundell, Neville King and Sir Thomas Clarges.

It is true that there was a good deal of land in South Lincolnshire owned by landlords who lived outside the area. But of these only the Earl of Bristol owned a really large estate. Institutions of various kinds owned some land, and they were rarely good landlords. Christ's Hospital owned most of the parish of Skellingthorpe and Guy's had several thousand acres in South Holland. St John's College, Cambridge, had owned much of the parish of Cranwell until the Thorolds had bought the land some time before the wars. A number of other colleges in Oxford and Cambridge owned land, as did the Church and the Crown, but the total acreage owned by outside landlords was not sufficient to explain the apparently minor role landlords played in improvement in South Lincolnshire before 1815.

A partial explanation may be found in the regional structure of landownership. Some 47 parishes were owned solely by one landlord; in a further 37 at least half the parish was owned by one landlord (Fig. 12). In this type of parish, there was more likelihood of a landlord promoting improvement than in a parish where property was divided among many different landlords. The parishes where property was concentrated in one or two hands were confined entirely to Kesteven. In Holland the squire was almost entirely

The Agricultural Revolution in South Lincolnshire

absent, and in every parish property was divided among many proprietors. In Gosberton, for example, there were 160 landowners, and in Quadring over 150. In Kesteven, on the other hand, property was much more concentrated except in two regions: the fenland

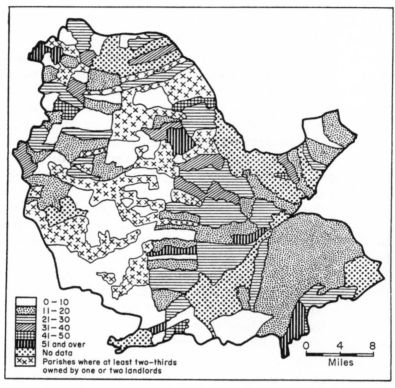

Fig. 12. The percentage of the Land Tax paid by occupier owners, by parishes, 1798–1808 (Holland 1798, Kesteven 1808).

and west Kesteven. But whereas in Holland *every* parish had divided ownership, in west Kesteven and the Kesteven fenland the parishes with greatly divided property were often next to parishes with only one landlord. Thus North Kyme was divided among 60 proprietors, but South Kyme was almost entirely owned by Sir A. Hume. In west Kesteven Sir John Thorold was the sole proprietor of Sedgebrook, but Long Bennington had over 110 landowners. West

Landlords, Farmers and Farms

Kesteven and the fenland were separated by a zone where there were few parishes with divided property; on the Heath and in south-east Kesteven the great landlord dominated farming to a greater extent than anywhere else in South Lincolnshire. The fact that the greatest advances were made on the Heath, and the least progress in west Kesteven and the fens, may be due to other factors; certainly in south-east Kesteven methods were poor, yet it was also an area owned largely by Ancaster, Whichcote, and Heathcote.

During the war period most farmers prospered. Not all the consequences of this prosperity were approved of by contemporaries. '...this period led to luxurious habits in families, and a seeking of trifling accomplishments for their children, quite anomalous to their probable spheres of life, and which would never otherwise have been thought of; thus frivolously occupying hours which ought to have been devoted to productive labour or to useful instruction'.[1]

Certainly many tenants must have preferred such conspicuous consumption to investment in new farming methods. Nor did a great many tenants use their wealth to buy their farms. In the seventeenth century occupier owners held a good proportion of English farmland, but they dwindled in importance throughout the eighteenth century. By 1785, according to one estimate based on the Land Tax returns for a number of Midland counties, they occupied no more than 11 per cent of the total farmland. However, the prosperity of the later eighteenth century, and particularly that of the wars, led some tenants to purchase their farms. By 1812, the height of the war boom, about 15 per cent of the farmland in these counties was owned by its occupiers.[2] Unfortunately it is impossible to trace the fortunes of occupier owners in South Lincolnshire over the same period, as the Land Tax returns for the area are incomplete. The only surviving returns for Holland are for 1798; returns for three wapentakes in Kesteven also exist for that year, but for the other six wapentakes the earliest returns are for 1808. There are complete returns for Kesteven in that year and also in 1812, but the

---

[1] Creasey, p. 364.
[2] Davies, p. 87; H. L. Gray, 'Yeoman Farming in Oxfordshire from the Sixteenth Century to the Nineteenth', *Quarterly Journal of Economics*, XXIV (Boston, 1910), 304-5.

The Agricultural Revolution in South Lincolnshire

TABLE 9. *Occupier owners*

A. A comparison of the Land Tax assessments, 1798 and 1812
Amount paid by occupier owners (to the nearest £)

Wapentake	1798	1812
Aveland	181	294
Bettisloe	39	81
Winnibriggs	42	75

B. A comparison of the Land Tax assessments, 1808 and 1812
Amount paid by occupier owners (to the nearest £)

Wapentake	1808	1812
Ness	320	320
Boothby Graffoe	162	327
Aswardhurden	256	277
Flaxwell	102	98
Loveden	385	475
Langoe	124	153

C. Number of parishes showing a gain, loss or no change in the amount of Land Tax paid by occupier owners, 1798, 1808 and 1812

	1798–1812	1808–12
Gain	20	36
No change	40	57
Loss	7	17

D. A comparison of the percentage of the Land Tax assessment paid by the occupier owners in 1798 with the percentage of the recorded acreage owned by occupiers in 1813

12 parishes increased
2 lost
2 showed no change

only guide to occupier owners in Holland at the end of the wars is the Court of Sewers Verdict for 1813.[1] The Verdict was an account of the acreage held by tenants and occupier owners in Kirton and Skirbeck Hundreds, and allows only an approximate comparison with the values recorded in the Land Tax returns of 1798.

The changes in the amount of Land Tax paid by occupier owners in each wapentake is shown in Table 9A, B. A more detailed

[1] Holland County Council Muniments Room, Boston, 32/1 and 8.

analysis is summarized in Table 9C showing the number of *parishes* which recorded an increase in the amount of Land Tax paid by occupier owners. Finally Table 9D illustrates the changes going on in Holland by comparing the percentage of the parochial assessment paid by occupier owners with the percentage of the acreage held by occupier owners in 1813.

Table 9 shows that there was a general increase in the importance of the occupier owner; only one wapentake showed a decline and this was small. But it was by no means a major change. In 1798 in three Kesteven wapentakes 3·6 per cent of the tax was paid by occupier owners but it rose only to 6·3 per cent in 1812. Similarly between 1808 and 1812 the amount paid in six wapentakes rose from 14 to 16·8 per cent of the total assessment. Most Holland parishes appear to have recorded an increase in occupier-owned land between 1798 and 1813 (Table 9D). However, the comparison of wapentakes perhaps strains the validity of the Land Tax returns,[1] a more reliable picture could be obtained by comparing parish figures. Table 9C shows that whilst a majority of parishes in Kesteven showed an increase between either 1798 and 1812, or 1808 and 1812, a surprising number showed a decline. There appear to be two reasons for this. In some cases occupier owners considered they would make greater profits from operating a larger farm, sold their own, and with the capital became tenants on larger farms. Secondly, in some cases enclosure compelled small landowners to sell their holdings to pay the costs. In Kesteven only sixteen enclosures took place at dates between surviving Land Tax returns. Of these, the eight parishes which intercommoned in Deeping Fen were enclosed in 1801; all showed a decline in the number of occupier owners between 1798 and 1808. The other eight parishes showed no decline, which is more in accord with recent research into the effect of enclosure on small landowners. It may be that the loss of grazing rights in Deeping Fen made many small farms there unviable units and compelled their sale.

The increase in occupier owners during the war period had little

---

[1] An examination of the validity of the Land Tax returns and an analysis of the South Lincolnshire documents can be found in D. B. Grigg, 'The Land Tax returns', *Agricultural History Review*, XI, part 2 (1963), pp. 82–94.

effect on the regional pattern, for the increases took place almost exclusively in the parishes where they had existed at the beginning of the wars. Although the proportion of occupier owners in South Lincolnshire as a whole was not much above the national average, they were largely concentrated into two regions. Fig. 12 (p. 84) shows the proportion of the Land Tax return in each parish paid by occupier owners; this gives an approximate indication of the percentage of the area farmed by them.[1] The regional differences correspond closely to the differences in landownership already noted, for the division of property in the fenland and in west Kesteven was, in fact, due to the prominence of occupier owners in these regions.

In a number of Holland parishes over half the farmers were occupier owners; generally, however, they occupied a smaller proportion than their numbers suggested. Only in a few parishes was over a third of the land farmed by them, although in no parish did this proportion fall below a tenth. In the Kesteven fenland occupier owners were numerous, particularly near Billinghay, but as noted earlier these parishes alternated with parishes where the occupier owner was quite absent. A similar state of affairs existed in west Kesteven where occupier owners were of above average importance, but were non-existent in some parishes. On the Heath and in southeast Kesteven a few parishes had a number of small freeholders, but only in two parishes did occupier owners farm more than a tenth of the land. Between the scarp of the Lincolnshire Limestone in the west and the fen in the east the land was farmed almost exclusively by tenants.

At the beginning of the nineteenth century the average farm in South Lincolnshire—whether occupied by tenant or occupier owner—was small, probably two-thirds being less than 50 acres (Table 11, p. 93). Furthermore, there is no indication that any substantial changes in farm size were taking place. This is contrary to the assumptions of a number of early writers on the agricultural revolution. According to H. Levy, for example, the small farm

---

[1] Neither home farms of substantial landlords nor the vicar's land have been counted as owner-occupied.

almost entirely disappeared in the century after 1750.[1] He argued that during this period prices for corn were relatively more favourable than those for livestock products. As the large farm was more advantageous for corn production than the small farm, there was a widespread amalgamation of small holdings and a steady increase in the average size of farms. The absorption of small farms was facilitated by enclosure, for the costs of enclosure were born by the landowners of each parish, and in many cases the smaller landowners were unable to meet their commitments; they were compelled to sell their holdings and larger farmers took this opportunity to expand their acreage.

Levy's theory has been challenged on a number of grounds. Studies of the Land Tax returns have shown that the decline of the small landowner had begun before the era of Parliamentary enclosure, and furthermore that enclosure was not necessarily followed by a decline in the number of occupier owners.[2] If the small farmer had been swept away he would have been of small importance in the statistics of farm size which are available in the Census of 1851 and the Board of Agriculture's returns of the 1860's. Yet these figures show that the farm of less than 50 acres was still by far the most numerous type of farm, as indeed it still is today. Finally, studies of estate surveys in the eighteenth and nineteenth centuries have revealed that whilst amalgamation was certainly going on, it was a slow process and by no means comparable with the almost catastrophic decline which Levy described.[3] It is now tacitly agreed that Levy overstated his case. The more cautious modern view has been well put by Professor T. S. Ashton: 'There is no reason why division and enclosure should have altered the acreage of the unit of control or the unit of production; but as the hand of custom was lifted, so the estates and farms could vary in size with the means and abilities of landlords and tenants, and generally the tendency was towards larger units'.[4]

[1] H. Levy, *Large and Small Holdings* (Cambridge, 1911), pp. 44, 45, 50, 51.
[2] A. H. Johnson, *The Disappearance of the Small Landowner* (London, 1909); J. D. Chambers, 'Enclosure and the small landowner', *Economic History Review*, x, no. ii (1940), 118–27; *idem*, 'Enclosure and the small landowner in Lindsey', *The Lincolnshire Historian*, no. 1 (Lincoln, 1947).
[3] *Census of Great Britain, 1851*, II, part 1, lxxx; G. E. Mingay, 'The size of farms in the eighteenth century', *Economic History Review*, 2nd series, XIV, no. iii (1962), 469–88.
[4] Ashton, p. 42.

Levy—and other writers who believed that enclosure swept away the small farm—drew their evidence largely from literary sources. Many eighteenth-century observers strongly deplored the social consequences of a decline in the number of small farms, and tended to exaggerate the extent to which this was happening. The people who really suffered from enclosure were the cottagers with a scrap of land and grazing rights on the common. In Lincolnshire it seems that Enclosure Commissioners and landlords made every effort to alleviate their lot, and the county was celebrated for its provision of cottages and land for these people.[1] But in any case even if this sort of person had been swept away by enclosure, it would have made little difference to farm-size structure. An excellent example of the situation is recorded by Thomas Hawkes. He wrote of Cowbitt Wash, near Spalding, '...a large and fertile tract of land and of great benefit to the most useful inhabitants of every country—I mean those little farmers and cottagers whose means of gaining honest livelihood is now become every day more difficult by the monopoly of such large tracts of land into the hands of a few overgrown individuals'.[2] Few of these 'farmers' would work more than an acre or two; the loss of grazing rights was certainly serious for them, but hardly justifies the picture of widespread absorption and amalgamation Hawkes's phrase conjures up.

Fortunately the existence of two surveys covering a considerable area allows some statistical evidence to be put forward. The Duke of Ancaster's large estate was surveyed in 1804 and again in 1830, and a complete list of his tenants and the size of their farms allows a comparison to be made. This affords a number of instances of amalgamation. In the parish of Laughton the duke's holding of 16 acres was held by four separate tenants in 1804 but had become one farm by 1830. In Osbournby in 1830 a farm of 31 acres had been split among three tenants in 1804; there are other examples of small units being amalgamated. But the whole estate covered over 20,000 acres; the overall changes which took place were relatively small, although there was a slight tendency for farms over 100 acres to

[1] W. H. Hosford, 'The enclosure of Sleaford', *Lincolnshire Architectural and Archaeological Society, Reports and Papers*, VII, no. 1 (Lincoln, 1957), 83–90; Young, *General View*, pp. 410–20; *Annals*, XXXVII (1801), 514–49.
[2] S.G.S./B.S./5/37.

TABLE 10. *The percentage of all holdings in each size group on the Duke of Ancaster's estate, 1804 and 1830*

	1804	1830
Farms of 300 acres and over	0·5	0·8
Farms of 100–299 acres	12·4	14·4
Farms of 50–99 acres	6·3	5·8
Farms of 5–49 acres	47·8	47·8
Farms of less than 5 acres	32·8	31·0

Source: L.R.O. 5 Anc 4.

increase in importance and those of less than 100 acres to decline (Table 10).

The fact that the small farm was still the dominant production unit at the beginning of the nineteenth century is perhaps the most weighty evidence that can be presented against the thesis that the small farm was being swept away. But the small farm was not of equal importance in every region in South Lincolnshire. Fig. 13 shows regional differences in the average size of farms. It is based on the number of occupiers recorded in the Land Tax returns, and consequently includes many holders of small amounts of land who were certainly not farmers. The average size will tend to be underestimated. As there are no accurate records of the amount of agricultural land in each parish during the wars, the early returns of the Board of Agriculture have been used. In most parishes the acreage of 1870 would have exceeded that of the 1800's, and this to some extent balances the inclusion of occupiers who were not farmers. There were very marked differences in farm size. The largest farms were to be found on the Heath and in south-east Kesteven. Farms in west Kesteven and the fenland were predominantly small. Within these three major divisions there were significant internal differences. The largest farms were on the Heath to the north of Sleaford and to the south of Grantham; in west Kesteven farms were generally slightly larger in the north of the plain, whilst in Holland the average size was slightly greater in the south than the north.[1]

The outline of farm sizes, indicated by the average size of farms,

[1] Creasey, p. 370.

The Agricultural Revolution in South Lincolnshire

is confirmed by the more detailed statistics on farm-size structure presented in Table 11. This table is based on a number of estate surveys made mainly in the first two decades of the nineteenth century, but also including one survey made in 1780 and another in 1830. The surveys cover 43,000 acres in Kesteven and

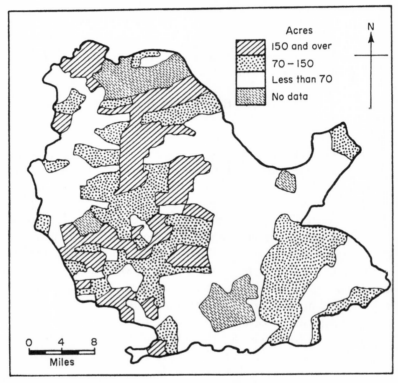

Fig. 13. The average size of holding, 1800.

46,000 acres of the two Holland Hundreds of Kirton and Skirbeck. The figures confirm the evidence already presented. The small farm was the dominant production unit in every region except the Heath to the north of Sleaford where farms of over 300 acres were the most important type; large farms were also of above average significance in west Kesteven and the southern Heath. Farms of less than 50 acres were three-quarters of all the farms in west Kesteven and the three fenland regions; in Kirton over four-fifths of the farms were of this size. In south-east Kesteven the medium-sized farm was of unusual importance, nearly a quarter of the farms being between 100 and 300 acres; this region was thus sharply distinguished from the Heath, for it had few large farms.

TABLE 11. *Regional differences in farm-size structure*

	Percentage of all holdings over 5 acres in each size group			
	300 acres and over	100–299 acres	50–99 acres	5–49 acres
The Heath (north)	40	10	20	30
The Heath (south)	10	14	12	62
West Kesteven	10	11	5	73
South-west Kesteven	2	24	14	60
Kesteven fenland	0	17	6	77
Kirton	0	4	9	86
Skirbeck	0·8	6·1	13·5	79
All Kesteven	5·3	19·6	10·2	64·8
All Holland	0·4	5·3	11·5	83·1

Sources: Marquis of Bristol's Estate Survey, *1780*; Sir John Thorold's Estate Survey, *1830*; L.A.O. Jarvis, 8, 2 Cragg 1/6, 2 Cragg 1/8, 2 Anc/uncat.; 5 Anc 4, Holywell/64/3; Holland C.C. Muniments Room, Boston, 32/1 and 8.

At first sight it might not appear difficult to account for these very marked regional differences in farm-size structure. For example, the two major areas of small farms correspond closely to the areas where occupier owners were of especial significance, and it is an implicit assumption of many writers that the occupier owner was usually a small farmer. Table 12 (p. 94) suggests that this is not so, for in Holland the proportion of tenanted and owner occupied farms in each size group in 1813 was much the same. More comprehensive figures are available, but at a much later date, for the whole of Kesteven and Holland. In 1913 although there was certainly a difference between the farm size structure for tenants and occupier owners, it is hardly sufficient to suggest that the presence of occupier owners accounts for a predominance of small farms in a region. A more likely explanation is the persistence of the custom of gavelkind.[1] Under this practice farms were not passed on intact to the

[1] Thirsk, p. 44.

The Agricultural Revolution in South Lincolnshire

TABLE 12. *Farm size and landownership, 1813 and 1913*

	Percentage of all farms in each size category					
	Holland 1813		Holland 1913		Kesteven 1913	
Size of farm	Tenants	Occupier owners	Tenants	Occupier owners	Tenants	Occupier owners
300 acres and over	0·3	0	5	4	12	12
150–299 acres	0·7	0·7	6	8	14	11
100–149 acres	3·7	2·0	6	4	9	7
50–99 acres	12·2	17·9	13	13	15	12
20–49 acres	26·0	24·3	27	21	19	21
5–19 acres	56·1	56·1	43	49	31	34

eldest son, but divided among all the heirs. In Holland this practice was still followed in the nineteenth century, and it may have been the custom in the past in west Kesteven.

Young and Stone both thought that the predominance of large farms on the Heath was due to the poor quality of the soil, but their arguments differed slightly.[1] According to Young the soils were so poor that only a farmer with abundant capital, on a large farm, could successfully work this land. Stone argued that because soil quality, and consequently the yields, were so low, a farmer had to have a large acreage to make a living. He went on to state that because of soil exhaustion, farms were getting larger. In the fens, on the other hand, it could be argued, the fertility of the soils was such that a reasonable living could be made from quite a small area. There was, of course, a close connexion between soil type and land use. It could be argued that because the Heath farms were predominantly arable, and because arable farming benefited from economies of scale, then farms on the Heath were large because they were arable. Similarly livestock farming could be most profitably carried out on small and medium-sized farms, and consequently the fenland and west Kesteven were areas of small farms. Yet whilst there is clearly a close relationship between soil, land use, landownership and farm size, the causal connexions are far from clear.

[1] Young, *General View*, p. 37; Stone, *General View*, p. 38.

CHAPTER VI

THE AGRICULTURAL REGIONS

There seems little doubt that there have been areas of specialized farming in England from an early date, but the idea of the agricultural region came much later. Indeed it was not until the County Reports made to the Board of Agriculture between 1793 and 1815 that the regional approach gained any widespread currency in agricultural writings. Half a century later a number of essays on the agriculture of English counties were published in the *Journal of the Royal Agricultural Society*. Many of these descriptions were regional in their approach, but as in the earlier County Reports agricultural regions were generally assumed to be identical with soil regions, which were somewhat inaccurately deduced from the geological maps of the time or from the observations of farmers and land agents in the area.[1]

Since the middle of the nineteenth century not only have the methods of regional delimitation become more sophisticated, but the Board of Agriculture's statistics (from 1866) have allowed quantitative approaches to replace the purely qualitative attempts of the nineteenth-century writers. The most important advance has been the recognition that a regional classification should not be defined in terms of the factors which cause regional differences, such as soil or climate; whilst soil type, for example, may in many cases be the prime determinant of regional variations in land use, it is not the only factor. Quite different types of farming may be found on identical soil types, whilst in areas of uniform soil type there may be a number of different types of farming. Nor is land use the sole criterion which distinguishes one farming region from another: farm size, farming methods, landownership and rent per acre may all differ significantly and contribute to the definition of

[1] H. C. Darby, 'Some early ideas on the agricultural regions of England', *Agricultural History Review*, II (1954), 30–47; W. G. Hoskyns, 'Regional farming in England', *Agricultural History Review*, II (1954), 3–11.

The Agricultural Revolution in South Lincolnshire

an agricultural region. Clearly these criteria are not necessarily a function of soil alone.[1]

The agricultural regions described here have as their basic unit the parish. Only two criteria were used to define the regions, the arable/grass ratio and the dominant crop combination. Clearly, other criteria could have been used, such as the predominant farm size, or the ratio of occupier owners to all occupiers. But the more the number of criteria used, the more complex the regions resulting, and in this chapter only a broad survey is attempted. Initially the parishes were divided into two classes, those which were mainly under grass and those which were mainly arable and rough grazing. As there are no reliable figures on the arable acreage at this time, this classification has necessarily been qualitative. All the parishes which were consistently described as 'mainly under grass' in contemporary writings were blocked out on a map. The remaining parishes were assumed to be either mainly arable land or rough grazing. The second step was to classify the arable parishes. The method adopted was a simplified version of J. C. Weaver's crop combination system.[2] The percentage of the total arable land in each parish under each crop was worked out, and the crops ranked in order of importance. In most parishes only three or four crops occupied any significant area; furthermore, these were found to occur in characteristic combinations. The diversity of soils and farm practice in South Lincolnshire meant that there were many crop combinations, but only four fundamental ones: barley and turnips, with wheat or oats; wheat, beans and barley; oats, rape and wheat, and finally oats, turnips and wheat. These four crop combinations, together with a number of less significant combinations were mapped, and the arable areas divided into regions on this basis.[3]

---

[1] D. Whittlesey, 'Major agricultural regions of the earth', *Annals of the Association of American Geographers*, XXVI (1936), 199–240; P. E. James and C. F. Jones (ed.), *American Geography; Inventory and Prospect* (Syracuse, N.Y., 1954), pp. 21–68, 259–71.

[2] J. C. Weaver, 'Changing patterns of cropland use in the Middle West', *Economic Geography*, XXX, no. 1 (Worcester, Mass., 1954), 15–47. For an example of this method applied to the 1801 returns, see D. Thomas, 'The acreage returns of 1801 for the Welsh borderland', *Transactions and Papers of the Institute of British Geographers*, no. 26 (1959), pp. 169–83.

[3] For a more detailed explanation and maps see D. B. Grigg, *Agricultural Change in South Lincolnshire, 1790–1875* (unpublished Ph.D. Thesis, Cambridge, 1961).

The Agricultural Regions

There were a number of significant differences in crop combinations within the grassland regions, and these were recognized as subregions. The major regions and the subregions are shown in Fig. 14. In Table 13, p. 98, the proportion of the total arable land under each crop in each region is shown.

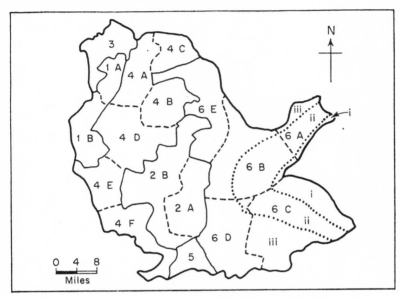

Fig. 14. The agricultural regions in 1801. 1A, West Kesteven (A); 1B, west Kesteven (B); 2A, south-east Kesteven (A); 2B, south-east Kesteven (B); 3, north-west Kesteven; 4, the Heath; 4A, the Cliff Row; 4B, Temple Bruer; 4C, north-east Kesteven; 4D, Ancaster; 4E, Grantham; 4F, Corby; 5, Deeping Fen; 6, the fenland; 6A, Skirbeck; 6B, Kirton; 6C, South Holland; 6D, Deeping Fen; 6E, Kesteven fenland; i, the marsh; ii, the Townland; iii, the interior fenland.

Thus, the parishes of the area can be grouped into a number of regions solely on the basis of land use. Unfortunately the parish has defects as a basic unit in an area as small as South Lincolnshire. Many parishes are so arranged that they include within their boundaries at least two quite contrasting types of land use; the aggregate figures for the parish then can be misleading. This is of particular significance in three areas. The row of parishes which straddled the Lincolnshire Limestone scarp between Lincoln and Grantham all contained limestone soils in their eastern portions and

The Agricultural Revolution in South Lincolnshire

TABLE 13. *The agricultural regions, 1801 (based on the 1801 crop returns)*

Region	Acreage of each crop as a percentage of total recorded acreage						
	Wheat	Barley	Oats	Potatoes	Peas	Beans	Turnips and rape
Mainly grassland							
West Kesteven (A)	29·8	22·4	19·8	—	—	5·5	19·8
West Kesteven (B)	24·5	30·0	13·4	—	—	18·7	11·8
South-east Kesteven (A)	17·0	25·5	33·5	—	—	8·8	14·3
South-east Kesteven (B)	31·4	18·2	23·9	1·9	3·9	5·7	14·7
South-east Kesteven (A and B)	25·1	22·5	26·6	1·3	2·0	7·4	14·9
Mainly arable							
North-west Kesteven	17·0	18·7	29·3	1·5	1·5	2·2	20·9
The Heath							
Cliff Row	18·5	38·8	8·8	0·4	6·5	1·3	25·4
Temple Bruer	41·7	26·4	10·3	2·1	3·1	—	15·8
North-east Kesteven	10·4	32·1	25·7	1·0	1·0	—	30·1
Ancaster	17·0	31·4	23·1	—	3·5	5·8	17·6
Grantham	13·2	36·0	19·6	—	—	5·8	22·7
Corby	27·7	34·6	16·6	—	—	1·7	6·8
Total heathland	22·2	33·2	16·6	0·6	2·8	4·1	20·3
Deeping Fen (Kesteven)	27·9	33·3	8·6	—	—	21·8	7·3
The fenland							
Skirbeck	27·2	11·4	40·0	4·8	—	6·0	10·9
Kirton	29·1	6·8	41·8	1·5	0·8	6·6	13·3
South Holland	28·0	6·8	52·5	1·4	3·3	3·8	7·1
Deeping Fen	6·7	5·6	82·8	1·1	—	0·9	3·1
Kesteven fenland	18·6	9·6	48·3	—	—	5·5	15·8
Total fenland	20·8	7·6	55·1	1·5	0·5	4·3	10·1

Lias clay in the west. In eastern Kesteven most of the parishes contained fenland as well as clays and gravels. But most important of all, the parishes in Holland nearly all stretched inland at right angles to the coast and consequently each included a section of Marsh, Townland and Interior Fenland. The fundamental regional division within the fenland, then, is not apparent at all in the statistics or the map. The fenland as a whole, is easily distinguished from the rest of South Lincolnshire (see Table 13); on the basis of the crop statistics, four subregions are recognized. In the regional description which follows, however, the differences between

The Agricultural Regions

TABLE 14. *The average rent per acre in the major agricultural regions*

	Assessment to the Property Tax, 1815	Arthur Young's estimates, 1799
North-west Kesteven	8s. 5d.	No data
West Kesteven	15s.	15s.
South-east Kesteven	11s.	15s. to 21s.
The Heath	7s. 3d.	8s. 4d.
Skirbeck Hundred	35s.	40s.
Kirton Hundred	31s.	40s.
South Holland	21s.	25s.

Source: B.P.P. XIX (1818), 235; Young, *General View*, pp. 49–52.

Marsh, Townland and Interior Fen are also discussed, but simply in qualitative terms.

Whilst the agricultural regions have been defined in terms of land use only, there was a broad correspondence between areas with a common soil type, a common form of landownership, predominant farm size and rent per acre (cf. Figs. 2, p. 15; 12, p. 84; 13, p. 92; and Table 14). Rent per acre is, in fact, a most useful criterion of regional differences. It is a function of a number of factors, but the most important are soil fertility, the type of farming, the size of holding and accessibility.[1] However, rent was also subject to a number of influences which have no regional significance. Landlord policy, for example, varied a great deal, but there seems little doubt that regional rent per acre is a most useful guide not only to differences in any one year, but also to regional changes over a period. In Table 14 the average rent per acre for the major farming regions in South Lincolnshire is given. Whilst the Property Tax assessments included the value of houses as well as land, they compare quite closely with Arthur Young's estimates.[2]

In the remainder of this chapter the major agricultural regions of the area are briefly described.

[1] D. R. Denman and V. F. Stewart, *Farm Rents* (London, 1959); on regional rents see Grigg, *Transactions and Papers*, no. 30.

[2] The rent per acre was calculated from the Property Tax assessments by dividing the total parish assessments by the total area of the parishes in each region. Young's estimates refer to an unspecified area and also to agricultural land alone, and so may be expected to be higher than those derived from the Property Tax.

The Agricultural Revolution in South Lincolnshire

WEST KESTEVEN

The grazing district of West Kesteven covered a greater area than is shown in Fig. 14 (p. 97) stretching westwards from the foot of the Marlstone scarp to the county boundary. About two-thirds of the agricultural land was under grass, which varied in quality, the better pastures occurring in the south. Both cattle and sheep were bred and fattened here, and the area also served as fattening pasture for livestock reared on the poorer grazing lands of the Heath. The arable land was partly used to supply fodder crops, but cash crops were also grown, notably wheat and barley. Malting barley found a ready market—except in wet years—among the maltsters of west Kesteven and Nottinghamshire. The main market was in Newark, but there were important breweries in Grantham and Lincoln, whilst maltsters were also to be found in a number of villages locally.

The most important feature of arable farming in this region was its backwardness. Many of the parishes were old-enclosed, probably to permit improvements in livestock farming; and the few surviving open-field parishes were enclosed early in the second half of the eighteenth century. The last open-field parish was Long Bennington, which was enclosed in 1796. Yet a visitor to Lincolnshire in 1800 stated that west Kesteven 'had more disagreeable features than any [area] we have hitherto examined'.[1] Even where parishes had been long enclosed the old open-field rotation of wheat, beans and fallow persisted. Turnips could only be successfully grown on the patches of gravel which most parishes contained; the area of gravels increased northwards, and turnips largely replaced beans as the main fodder crop, allowing two subregions to be distinguished (Table 13, p. 98). The clays were largely without underdrainage, and this not only reduced the quality of the wheat, and particularly the barley crop, in a wet year, but encouraged foot rot among the large sheep flocks.[2] In addition, the Witham and the Brant, which meandered across much of this region, very frequently overflowed. This was due to the narrowing of the Witham in its passage through Lincoln and to obstructions at Brayford Head in the town.[3]

[1] *The Farmer's Magazine*, I (1800), 394. [2] Thirsk, pp. 301–2.
[3] *Annals*, XXXVII (1801), 536; L. A. O. Cragg 1/1.

The Agricultural Regions

Nevertheless, the clay loams of this region had a considerable natural fertility. It was not difficult to maintain the organic content of the soils and excellent yields of wheat and barley (when limed) were obtained in a normal year. But the heavy texture of the clays meant that cultivation was expensive, and the region was essentially a high-cost wheat producer. The majority of farmers kept the bulk of their land under grass. Even if their pastures were inferior to those of either south-east Kesteven or the Townlands, they were saved the high production costs of arable cultivation. In eighteenth-century Lincolnshire grassland, even of very moderate quality, bore a higher rent than all but the very best arable;[1] this premium, together with the high yields which could be obtained, helps to account for the high rent per acre in the region. Only the fenland bore higher rents in 1815 (Table 14, p. 99).

The predominance of the small farm may also help to explain the high level of rents. Although the region had a greater proportion of large farms than many other parts of South Lincolnshire, the medium-sized farm was rare, and the small farm was by far the most important size group. In a number of parishes over a fifth of the farm land was occupied by its owners. Indeed not only may the predominance of small farmers and the relative importance of occupier owners have helped account for the high rent per acre to be found here, but their presence may also have been a significant factor in the low level of farming methods. This was a region where there had been a noticeable increase in the number of owner occupiers during the wars; neither they nor the large number of small tenant farmers had the capital to risk new methods. The experiences on the poorly drained clays probably discouraged the few who had been enterprising.

A last factor to be noticed was the excellent position of the region. Three large market towns—Newark, Lincoln and Grantham—were all within fairly easy reach; combined with the inherent fertility of the soil, this had long made the region favourable to farming, and the presence of large nucleated villages, a sharp contrast to the sparsely populated Heath to the east, reflected a long-sustained

[1] *Essays on Agriculture, occasioned by reading Mr Stone, A Report on the County of Lincoln, by A Native of that County* (London, 1796), p. 47.

prosperity. But the clays were essentially high-cost producers; the necessity for underdrainage and the high cost of cultivating the soil meant the high yields were accompanied by high production costs. More significant, the clays were poorly suited to the new farming methods; the turnip could not easily be grown and the folding of sheep was a risky matter. The prosperity of the region was, in fact, coming to an end; whilst good profits were undoubtedly being made here during the war, the region was not developing as rapidly as the Heath or the fenland. Records of the receipts to Schedule B of the War Income Tax survive for 1806 and 1815 for each of the parishes of South Lincolnshire.[1] A comparison shows not only that west Kesteven generally had a lower rate of increase than either the Heath or the fen, but that the southern parishes of west Kesteven were increasing less than the northern parishes, where there were more extensive areas of gravels; here land was more easily worked and the folding of sheep on turnips possible and profitable (see Table 15).

TABLE 15. *Regional differentiation in rent per acre*

Region	1806–15 (% increase)	1815	1842–3	1859–60
The Heath	56	100	100	100
North-west Kesteven	80	110	100	105
South-east Kesteven	36	150	110	102
West Kesteven	37	190	140	120
The fenland	38	430	210	180
(Skirbeck	94	—	—	—
Kirton	33	—	—	—
South Holland)	30	—	—	—

In 1815, 1842 and 1860 the lowest average rent—in each year the Heath—is represented as 100 and the other regions proportionally.

Sources: *B.P.P.* xxxii (1814), 451 ff.; *B.P.P.* xxxix (1860), 157 ff.; P.R.O./E. 182/571–587.

NORTH-WEST KESTEVEN

Northwards along the low plain of west Kesteven there was a very marked change in land use, and a separate region, north-west Kesteven, can be distinguished. Pasture rapidly declined in im-

[1] Schedule B was a tax on farmers' profit. The difficulty of estimating this was recognized and the assessment was taken as a proportion of the rent (A. Hope-Jones, *The Income Tax in the Napoleonic Wars*, Cambridge, 1939, p. 20).

The Agricultural Regions

portance, rough grazing became more extensive, and the crop combination changed. A group of parishes to the west and south-west of Lincoln were the only parishes in South Lincolnshire still to grow rye, whilst the major crop was oats. These changes reflected a change in soil type, for here the Lias clays are largely concealed by gravels, giving an infertile, but easily tilled and drained, sandy soil. There was more woodland here than to the south and much of the area was still under gorse or poor grazing. On the other hand this was a region where the new methods could easily be adopted, for turnips grew well and sheep were in no danger of foot rot. Crop yields were low, but the soils responded better to fertilizers than did the cold clays, and there were no difficulties with underdrainage. Nor did the open fields survive in many parishes, for this was a region which had largely been enclosed before the era of Parliamentary enclosure.[1] It seems to have been a region where farming standards were low but greater progress was being made than in west Kesteven.

Nevertheless, even at the end of the war, rent per acre was lower than anywhere else in South Lincolnshire (Table 14, p. 99). Few parishes in north-west Kesteven had an average rent of more than 10s. an acre, and the highest rents in north-west Kesteven were to be found on the clays. The low rents were almost entirely due to the poor soil, for the region was near to Lincoln, and there were a great many small farmers. The uniformity of land use was not matched by the pattern of land ownership. Parishes with a predominance of small farms lay next to parishes with a few large farms; a few parishes had a remarkable number of occupier owners, whilst in contrast others were in the sole ownership of a squire.

THE HEATH

The Heath stretched south from Lincoln through Ancaster to Grantham, and then south-eastwards towards Stamford. It was predominantly under arable, but there were still extensive areas of rough grazing and some waste. Indeed bad farming had in some places resulted in a reversion to waste or warren during the latter

[1] D. R. Mills, 'Enclosure in Kesteven', *Agricultural History Review*, VII, part ii (1959), 83–7.

part of the war period. Not surprisingly in such a large region there were significant local variations in crop combinations, but the overall pattern was dominated by barley, wheat and turnips. In the parishes of the Cliff Row (Table 13, p. 98) the Norfolk system seems to have been imported *en bloc*. The four course was followed, and sheep were folded on the turnips. East and south of this sub-region the crop combination changed. In north-east Kesteven the Heath parishes included a substantial acreage of fen, and this is reflected in the importance of oats and rape. Southwards around Temple Bruer there was a further change, wheat replacing barley as the main crop. This was almost certainly due to the persistence of open fields here. Enclosure, among other advantages, invariably allowed a closer adaptability of crop to soil type; before enclosure wheat was often grown regardless of whether the soil was suitable for it or not. South again from the Temple Bruer area oats became the leading crop; this was due partly to the decline in soil fertility south of the Ancaster Gap and also to the fact that a considerable proportion of the land had recently been ploughed for the first time. Under these circumstances oats was invariably the first grain grown after the original turf had been pared and burned.

The quality of farming varied as much as the crop combinations. The nature of farming practice here had been a matter of some controversy between Young and Stone. Young had visited the Heath in 1770 when there was little but gorse and rabbit warren. He was greatly impressed by the changes which had taken place when he returned in 1799, brought about mainly by enclosure.

The vast benefit of inclosing, upon inferior soils, can rarely be seen in a more advantageous light, than upon Lincoln Heath. I found a large range, which formerly was covered with heath and gorse, yielding in fact little or no produce, converted by inclosure to profitable arable farms; let on an average at 10/- an acre; and very extensive country all studded with new farm houses, barns and offices, and every appearance of thriving industry; nor is the extent small, for these heaths extend nearly seventy miles; and the progress is so great in twenty years that very little remains to do.[1]

This was undoubtedly a much exaggerated picture. Young's itinerary suggests he hardly went on the Heath proper at all south

[1] Young, *General View*, p. 77.

The Agricultural Regions

of Lincoln; and his description hardly tallies with other descriptions including Stone's. When Young visited Lincolnshire the few remaining open fields in South Lincolnshire were all, with a few exceptions, on the Heath. The excellent farming described by him was certainly to be found among some farmers, especially near Lincoln, but the general practices were most primitive, so that land in a number of parishes had to be returned to rabbit warren. In fact the Heath had been reclaimed before sufficient farmers had the knowledge necessary for its successful cultivation. If it was to be farmed at all it had to be farmed well, and the crux of the problem was to increase and then maintain the organic content of the soil. The Norfolk system was admirably suited to this purpose, and conversely the soils were suitable for the system. Turnips could be grown and sheep easily folded. Although yields were low, there were no problems of underdrainage, and the thin soils were easily tilled. The Heath presented a striking contrast with west Kesteven, for although rents per acre were still much lower than anywhere else, except in north-west Kesteven, they were increasing rapidly throughout the war period.

Not only were the soils of the region conducive to the adoption of the new methods but the pattern of farm size and landownership was also favourable to improvement. Occupier owners were rare and many of the parishes were owned by one landlord alone. Many of the great landlords—Thorold, Turnor and Welby in particular—had part of their estates in this region. Farms were generally large, particularly to the north of Sleaford and immediately to the south of Grantham. Both factors may have favoured the more rapid spread of new techniques in this region in comparison with either the fenland or west Kesteven. The size of farms also affected rent per acre. Agricultural rent is paid partly for the land, partly for the buildings. As the number of buildings necessary does not increase proportionately with the increase in acreage, the rent per acre of a small farm, all other things being equal, will be greater than that of a large farm.

SOUTH-EAST KESTEVEN

From Ewerby, just to the east of Sleaford, southwards to Edenham and south-west to Bassingthorpe, the greater part of the agricultural area was grassland. In a number of parishes as much as 90 per cent of the land was under grass, although the average regional figure was certainly much lower. This was a region—south-east Kesteven—of graziers, and arable farming was quite subsidiary to the breeding and fattening of cattle and sheep. Cattle were imported for finishing from many parts of England, Scotland and Wales. Sheep, which were much more numerous, were largely locally bred, though some were brought from the nearby Heath to be fattened. The small amount of arable land had a variety of uses. In the region as a whole wheat was the major crop, but there were considerable variations. The easternmost parishes each contained fenland, where oats and rape were the major crops. Elsewhere the boulder clay cover had been partly breached by erosion and the underlying limestone revealed. Thus, within each parish, barley and turnips were grown on the limestone—or 'creech' land, as it was locally called—whilst wheat and beans were found on the clays.[1]

On the basis of cropping then, a distinction can be made between the eastern row of fenland parishes, and the remaining parishes which were generally on clays. This division was also apparent in landownership and farm size. Many—but not all—of the fenland parishes had a high proportion of land worked by occupier owners, whilst to the west there were virtually none. Indeed, much of this area was owned by the Duke of Ancaster, whilst Sir Thomas Whichcote and Sir Gilbert Heathcote also had substantial estates here. There was rather more uniformity of farm size, for in both the fenland subregion and to the west the medium-sized farm predominated; few farms were above 300 acres, whilst the small farm was of less significance than in either west Kesteven or the fenland proper. There were other differences. The Heath is gently undulating in relief and was primarily enclosed during the later eighteenth century. In south-east Kesteven enclosure had been largely achieved by 1700. The limestones have been covered by boulder clay, into which a number of rivers have cut deeply, giving a much more broken relief, with small, hedged fields and winding roads. On the Heath roads run straight for considerable distances and stone walls are as common as hedges. Indeed this difference in topography and date

[1] L.A.O. Cragg 1/1. See also the Tithe Redemption Awards in Lincoln Archives Office.

The Agricultural Regions

of enclosure may help to explain the difference in farm size. It is often held that rugged topography prevents the building up of large farms; certainly the large farms of England are mainly to be found on areas of level terrain.

Just as the farms were mainly intermediate in size, so were rents. In spite of the high proportion of grazing land in the region, rent per acre was below that of west Kesteven and much below that in the fenland. This was due to a number of reasons. The small amounts of arable land were of very poor quality and gave poor yields. The villages of the area were generally small, and had very bad transport connexions with the market towns, which with the exception of Corby were outside the region (see Fig. 6, p. 42). Lastly, because of the predominance of grazing, the improvements of the period—essentially in arable farming—had little impact on the region. Most of it had been enclosed long before the era of Parliamentary enclosure, and so one form of stimulus to change was lacking. As a consequence rents showed only a small percentage increase during the wars in comparison with the Heath (Table 15, p. 102), although rent per acre remained in 1815 above the rent of that region.

THE FENLAND

The Fenland is easily distinguishable from the rest of South Lincolnshire, not only by differences in crop combinations, but in soil type, relief and a variety of other criteria. There was, however, one transitional zone which merits attention before the fenland proper is considered, and that is the group of parishes around Market Deeping. The main crops here were wheat, barley and either peas or beans (Table 13, p. 98). Before enclosure in 1801 these parishes had all intercommoned in Deeping Fen, but unlike any of the Holland parishes which practised intercommoning, the Kesteven villages also had open fields. These fields lay on an area of fertile gravel soils on the edge of the fen proper.[1] Whilst this area thus differed from the fenland in its crop selection it was very similar in other ways. The farms were predominantly small and there were many occupier owners. It was probably the extreme division of property which had delayed enclosure so long; and it was almost

[1] This area has been aptly described as pseudo-fen (Stamp, part 76 and 77, p. 512).

The Agricultural Revolution in South Lincolnshire

certainly the persistence of the open fields which accounted for the growing of beans rather than turnips on the light sandy soils.

So good were these soils that in the half century preceding enclosure the open-field regulations were increasingly relaxed and a bare fallow dispensed with. The parishes also had abundant grazing resources in Deeping Fen—although it appears that these lands were being overstocked towards the end of the century—and a good location hear the market towns of Spalding and Stamford. Not surprisingly, rent per acre was much higher than on the Heath to the west or on the undrained fen to the east. Enclosure in 1801 improved the prospects of the region, and rents rose steadily between 1806 and 1815.

The Fenland was distinguished from the rest of South Lincolnshire by the dominance of oats and the cultivation of rape, a crop hardly grown elsewhere. In the Fenland as a whole oats occupied 55 per cent of the arable land; in Deeping Fen four-fifths of the land was under the crop, and in no subregion did the proportion fall below 40 per cent. In comparison rape was of small importance, occupying no more than a tenth of the arable acreage. Indeed the only significant crop other than oats was wheat, covering a fifth of the total recorded arable. Farms throughout the fenland were much smaller than in the rest of South Lincolnshire, nearly four-fifths being under 50 acres, whilst the occupier owners were more numerous and farmed a greater proportion of each parish. The fenland was sharply delimited to the west not only by soil but in relief; and it was in fact the flatness and lack of gradient which gave rise to the problem of drainage, the major preoccupation of most fenland farmers.

But not surprisingly in such a large area there were local variations in farming practices. Tables 13 and 14 (pp. 98 and 99) show there were differences between the south and the north, and these were directly related to drainage efficiency. North of Deeping Fen the drainage of Kirton and Skirbeck was more successful than the schemes in Deeping Fen and South Holland; this is reflected in the decline of rent per acre from north to south and also the changes in crop combinations. Where drainage had been successful wheat could be grown and there were in fact two parishes in Skirbeck

The Agricultural Regions

Hundred where wheat was the leading crop. In the less efficiently drained areas spring-sown oats remained the leading crop.

The dominant regional differences within the fenland, however, were not between the north and the south but between the Marsh, the Townland and the Interior Fen. Unfortunately these cannot be discussed on a quantitative basis, but the differences are clear from contemporary descriptions.

THE MARSH

The Marsh consisted of land which had been enclosed from the Wash over a very long period; the width of the Marsh depended partly on the shape of the coastline and the direction of winds and currents, and partly on the progress of embankment and enclosure. It was widest in South Holland, where the last great enclosure was undertaken in 1793. The soils were generally light and sandy silts, and near the sea were salty, being of little use other than for sheep grazing. Farther inland they could be cultivated, and throughout the wars land was steadily ploughed up. Oats was the main crop, but a wide range of crops was possible; this was one of the few parts of the fenland where turnips were successfully cultivated. Farming practices were generally deplorable, and as late as 1820 the Marsh farmers in Skirbeck Hundred were taking ten successive grain crops without fallowing. The Marsh was, on the whole, free from drainage difficulties; small drains led into the Wash through sluices in the coastal banks. However, floods from the sea were an ever present danger, and in 1811 the whole of the Marsh was drowned when the embankments in Kirton and Skirbeck were breached.

THE TOWNLANDS

The Townlands were the centre of original settlement; standing slightly above the interior fenlands they were beyond the reach of all but the most serious floods. At the end of the eighteenth century this region was a grazing district whose pastures could be matched in quality and rent in few parts of England. Before the enclosure of the Interior Fens, Townland farmers had grazing rights in the common fen. Enclosure led to some consolidation of holdings and the rise of independent farms and settlement in the interior, but the practice of holding lots in the fen persisted until well into the nineteenth century.[1] Where the Townlands had been ploughed a variety of crops were grown; the heavier silts favoured the predominance of wheat and beans, but oats and rape were also grown. A fallow was common, not as a residual feature of the open fields, for they had long since disappeared from this region, but as an essential part of clay-land farming. Both beef cattle and sheep were fattened here, but there was relatively little breeding. Sheep were brought in from the Heath and the Wolds, and indeed many Wolds farmers actually owned land on the Townland to which they sent their sheep to be finished. Rent per acre on the Townlands was higher than anywhere else in South Lincolnshire, for not only were the pastures good, but the arable was fertile and farms were small. Unfortunately the quality of the land was not matched by the standard of the farming, for few graziers had experience of arable cultivation, and the fertility of the soils led them to ignore the most elementary principles of good husbandry.

THE INTERIOR FENS

The Interior Fens differed strikingly from the Townlands and the Marsh. Until the later eighteenth century this area had been little more than an appendage to farms on the Townlands or on the 'highlands' of the Kesteven fen parishes. But enclosure in the later eighteenth century not only led to an improvement in the drainage but to an increase in arable land and to some new settlement. Land use varied a great deal during the wars, the state of the drainage being the decisive factor. As a rough generalization there was more arable in the northern fenland than in the south; the longer the drainage has been in existence, the more likely was there to be wheat as well as oats. Thus in the southern fenland oats were a remarkably high proportion of the arable acreage, rape being often the only other crop. In the north more wheat was grown, although only in a few areas did it exceed the oats acreage. Summer grazing remained a major form of land use, and grassland was only ploughed up when

[1] G. J. Fuller, 'Development of drainage, agriculture and settlement in the fens of South-east Lincolnshire during the nineteenth century', *East Midland Geographer*, no. 7 (Nottingham, June, 1957), pp. 3–15. See also L.A.O., Tithe Redemption Awards.

The Agricultural Regions

a combination of dry years and high prices for grain encouraged fen farmers to face the risks of a drowned harvest. The Interior Fens remained, in fact, to the end of the wars, 'half yearly' lands; this is well shown by the difference in rent per acre between Kirton and South Holland (Table 14, p. 99).

This brief summary of the agricultural regions illustrates three important characteristics of farming during the Napoleonic Wars. First of all, there was a close relationship between soil type and land use; secondly, there was a high degree of regional differentiation; and thirdly, farming was essentially extensive and technologically backward.

Although the agricultural regions which have been discussed in this chapter were based on elements of the agricultural landscape rather than simply on the soil regions of South Lincolnshire, it is clear that in fact there was a very close relationship between soil type and land use. Thus the three grazing districts were all on heavy clays; in crop selection there was an even closer connexion. The characteristic crops of the limestone Heath were barley and turnips, of the clays wheat and beans, and on the peat soils of the fen oats and rape. Within some of the major regions there was an even closer adjustment to soil type. Thus in south-east Kesteven wheat and beans were grown on the boulder clays, barley and turnips on the 'creech' soils. In west Kesteven gravels occurred in nearly every parish, although clay loams were the predominant soil type, and it was mainly on these soils that turnips were grown. The close adjustment between soil type and crop selection had been made possible by enclosure, for on the open fields custom still determined the rotation. Thus of the three small regions in South Lincolnshire where there were noticeable failures to adjust cropping to soil type, two were where open fields persisted; around Market Deeping and to the north of Sleaford (Figs. 7, 10, pp. 51, 73). In the third region, west Kesteven, barley was grown in surprisingly large acreages on clays; this can be attributed to the proximity of the Newark breweries.

As long as farming methods remained backward, farmers were severely limited in crop selection by the inherent characteristics of

The Agricultural Revolution in South Lincolnshire

their soils. Thus on the limestones, where there was little manuring, wheat could be grown, but it would yield below the average. On the other hand barley thrived better. The reverse was true on the clays. It obviously payed a farmer to select the crops which experience showed did best on his soils. Later in the nineteenth century the spread of artificial fertilizers meant that a farmer could supply the chemical deficiencies of his soils, and thus widen the range of his crops. But in 1801 not only did soil type influence crop selection, but it also determined overall productivity. Where neither artificial fertilizers or farmyard manure are used in any quantity, crop yields are largely a function of the inherent nitrogen content of the soil. Under these circumstances the fen and clay soils had a considerable advantage over the gravel and limestone areas; this is clearly demonstrated in the wheat yields of these regions (Fig. 9, p. 61).

The second characteristic was the striking degree of regional differentiation. Thus three regions—the Townland, west Kesteven and south Kesteven—had between 65 and 80 per cent of their area under grass, whilst the remaining parts of South Lincolnshire were mainly arable or rough grazing. The great increase of arable during the wars exaggerated this contrast, for there was little ploughing of the permanent grassland whilst much of the rough grazing in the arable regions was converted to arable. Contrasts in crop combinations, as has been seen, were equally marked. But regional differentiation was not confined to land use. Contrasts in farm size, landownership, rent per acre and enclosure history were all considerable. Regional variations in rent per acre were greater than at any time in the nineteenth century, and this reflected not only contrasts on soil fertility and the state of drainage, but also in farm methods. Table 15 (p. 102) is an attempt to show the degree of regional differentiation in 1815, 1842 and 1860. In each year the region with the lowest average rent—the Heath—is represented by 100, and the rent of other regions shows proportionally. The difference between the highest rented region and the lowest rented region was greater in 1815 than in 1842 or 1860.

Although differences in landownership, farm size and enclosure history followed approximately the divisions of soil type, the precise

The Agricultural Regions

relationship is by no means clear. Differences in farm size may have reflected contrasts in the type of farming, which ultimately can be traced back to soil type, but it seems that features such as the persistence of gavelkind were as important. Although the distribution of occupier owners shows a close correspondence with the area of better soils—where, it could be argued, the small landowner could hold his own more easily than on the poor Heath soils—there is no unequivocal evidence that the distribution was related to this factor. Enclosure history again shows some evidence of the influence of soil type and it is often argued that early enclosure occurred in those areas most suitable to livestock farming. In South Lincolnshire these areas corresponded to the heavier clays, which were not suitable for arable farming in terms of the then technology, and furthermore gave a good pasture. Certainly the three regions of grassland were predominantly old-enclosed, but there were enough surviving open-field parishes in the grazing districts to prevent a wholly valid generalization being made about the connexion between soil type and the date of enclosure. Clearly other factors affected the rate and timing of enclosure.

Nevertheless, it is difficult to deny the high degree of regional differentiation and the broad correspondence of agricultural regions with soil regions; equally important was the marked difference in the suitability of these regions for the New Husbandry; and thus the rate of agricultural change varied too.

Finally it should be noticed that farming remained essentially extensive, in spite of the specialization of regional farming which had emerged by the time of the wars. Intensity in farming is essentially a relative term, relating labour and capital inputs to land; but in no region by the end of the wars could farming be said to be intensive. Part of the explanation of this was the continued presence of rough grazing; farmers who wished to increase their output had no need to improve their farming methods—or increase the intensity of farming—whilst there was land left to be ploughed up. A further explanation was the failure of all but a few farmers to appreciate the underlying ideas of the New Husbandry. The Norfolk system and its variants, by using arable land to provide fodder crops, by increasing labour and capital inputs, and by elegantly interlocking arable

The Agricultural Revolution in South Lincolnshire

and livestock farming on the same farm to the benefit of both enterprises, necessitated an intensification of farming and a radical change of land use. In the thirty years after the end of the wars, this 'mixed' farming, a concept known to but a few in 1815, spread to every part of South Lincolnshire; and there followed as a consequence a great increase in the intensity of farming, a replacement of non-integrated, specialized farming by interlocking arable and livestock husbandry, and as a consequence changes in land use which led to a decline in regional differentiation.

PART II
AGRICULTURE AFTER 1815

CHAPTER VII

PARADOX AND PROGRESS

The 50 years which followed the end of the Napoleonic Wars saw fundamental changes in the farming of South Lincolnshire. New methods were adopted much more generally and more rapidly than in the preceding half century, and as a result wheat yields nearly doubled in the space of 30 years. Whilst much the same range of products was produced in 1850 as in 1815, the manner in which they were produced was very different. Mixed farming, with a greater proportion of the land under arable, and particularly under fodder crops, replaced both the specialized grazing and the backward arable farming of the war period, resulting in a substantial increase in the total arable acreage. As late as 1815 South Lincolnshire was still considered to be primarily a grazing country, but by 1850 grassland was much less common and by 1875 less than a quarter of the agricultural land was pasture.

Yet these striking increases in the arable acreage and productivity took place in a period apparently unfavourable to farming and to grain production in particular, for the year 1815 brought not only peace but the end of a generation of prosperity for English farmers. Indeed the good times had come to an end before the wars were over. Wheat prices had reached an astonishing peak in 1812 but they had been halved by 1815. Nor was the dislocation brought about by the end of the wars temporary. Farmers in 1815 and 1816 were sure that prices would return, if not to the peak of 1812, at least to a fairly high level, and this optimism persisted throughout the 1820's and 1830's. These hopes, however, were never justified, for the long-term trend in wheat prices was downwards from 1812 (see Fig. 15).[1]

[1] Fig. 15 is based on Lord Ernle, *English Farming Past and Present* (London, 1932), pp. 488–9; B.P.P. XVI (1895), 256, LXV (1878–79), 425, 431. General discussions of the post-Napoleonic depression can be found in L. P. Adams, *Agricultural Depression and Farm Relief in England, 1813–52* (London, 1932); G. E. Fussell and M. Compton, 'Agricultural adjustments after the Napoleonic Wars', *Economic History*, III, no. 14 (1939), 184–204; G. E. Fussell, 'Agricultural depression a century ago', *Journal of the Land Agents Society*, London, 1938; Ernle, pp. 316–31; Clapham, II, 133–42.

The Agricultural Revolution in South Lincolnshire

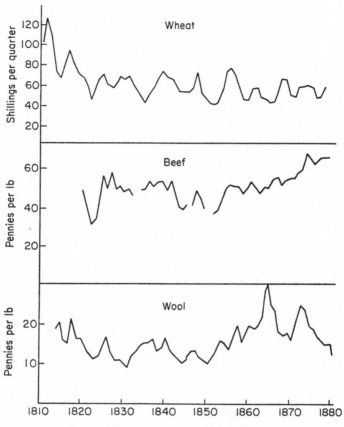

Fig. 15. Wheat, beef and wool prices, 1811–80.

The farming community greeted this with a mixture of exasperation and desperation, which is well illustrated by a letter a Doddington landlord wrote to his agent in 1836.

With regard to your suggestion about a further abatement of rent, I do not see how it can be avoided. The price of wheat is so much lower than the most thorough going enthusiast for its reduction could have brought himself to believe in the wildest of his speculations at the close of the French War. Eighty shillings was then thought but barely a remuneration. Sixty shillings a few years later was thought just enough to save, and now

Paradox and Progress

that we can scarcely get what was then supposed to be the lowest price at which it could be imported we are still urged to throw the market open in order to import it lower from countries not paying our taxes or poor rates.[1]

Colonel Jarvis was right in believing that the increased taxation of the war period which had been carried over into peacetime, was a heavy burden to the farming community; he neglected, however, the fact that the general price level had also fallen, making the increasing number of producer goods which a farmer purchased cheaper. But this was not enough to compensate for the low prices.[2] The Corn Laws were revised in 1815 and 1828, but this did not stem the decline. Not surprisingly South Lincolnshire farmers felt strongly about the repeal of the Corn Laws and in the early 1840's there were many gloomy forebodings. 'What the effects...may have I cannot tell,' wrote a tenant of Lord Willoughby's in 1842, 'but the prospects seem anything but encouraging to occupiers of English soils.'[3] Prices did not fall immediately after repeal, but between 1848 and 1852 wheat prices were particularly low, and many farmers—as before and since—blamed the government. Mr Stevenson, who farmed in Skellingthorpe, saw little prospect of prosperity in 1850.

It is evident the present Government intend to persist in Free Trade in Corn and it is equally evident in my mind that from the immense importations we continue to receive...that the British Farmer cannot at all compete with the Foreigner even in average years, but in years like the present, when we have little more than half crops only, and those of most inferior quality, the loss must be dreadful to contemplate and many thousands of the industrious present race swept away. I propose applying every industry and economy for one year more when if I cannot make a living I wish to restore my farm into the Governor's hands.[4]

If Mr Stevenson had been a little more persistent he would have 'made a living' in the next few years. Wheat prices rose during the Crimean War, and in the 20 years afterwards, whilst they did not continue to increase, were less susceptible to the marked cyclical fluctuations which had presented such problems to farmers between

[1] L.A.O. Jarvis 8.
[2] W. T. Layton, 'Wheat prices and the World's Production', *J.R.A.S.E.* LXX (1909), 102; Adams, pp. 33-4.
[3] L.A.O. 2 Anc 7/10/45. [4] L.A.O. T.L.E. 38/7/212.

The Agricultural Revolution in South Lincolnshire

1815 and 1852. Indeed it was the uncertainty almost as much as the lowness of prices in these decades which caused distress. After the initial break in prices in 1814, wheat recovered to 1818, but fell again to a particularly low point in 1822. Prices were stable in the remainder of that decade, but a run of good harvests in the early 1830's depressed the market again. The recovery which coincided fortuitously with Queen Victoria's accession did not last long and in the late 1840's there was again a period of low prices.

The prices of oats and barley followed approximately the same long-term trend as that of wheat,[1] with, however, two important exceptions. Good malting barley suffered smaller proportional falls in price throughout the period and was less subject to violent fluctuations. Further, in the 1830's, when wheat was falling rapidly, barley prices were rising. Livestock prices were maintained much better than any of the grains. Thus whilst the price of animals sold for beef at Smithfield did show a downward trend between 1820 and 1852 (Fig. 15, p. 118), the decline was much less than for grain crops. Wool prices also fared better, and Lincolnshire wool in particular, for it fetched a higher price in Bradford than most other native wools. Wool prices held up well until about 1820, and it was not until the middle 1820's that graziers in South Lincolnshire suffered any great distress. In the 1830's wool prices rose steadily, in contrast to wheat, but fell again in the 1840's. After 1852, however, prices for livestock products sharply diverged from those for corn. The average annual price for Lincolnshire Longwool in 1865-9 was 75 per cent above that in 1845-9, whilst between 1848-53 and 1874-5 there was a 33 per cent increase in beef prices and a 45 per cent increase in mutton.[2]

Thus, whilst the years following the Crimean War were generally prosperous, the 1820's, 1830's and 1840's were a less happy period for English farmers. Nor is there any shortage of literary evidence of the distress. In 1816 the Board of Agriculture published the replies to a Circular Inquiry made on agricultural depression, whilst Select Committees of the House of Commons investigated conditions in 1821, 1822, 1833 and 1836. In the latter year a Select Committee

[1] J. A. Venn, *The Foundations of Agricultural Economics* (London, 1933), p. 387.
[2] *J.R.A.S.E.* XXXIX (1878), 475-6.

also reported to the House of Lords. From this evidence, together with the letters written in these years by farmers and landlords in South Lincolnshire, it is evident that there was considerable distress in the area, and that all regions suffered at some time between 1815 and 1852.

The initial break in prices at the end of the wars, not surprisingly after a long run of prosperity, brought widespread distress. Amongst the worst sufferers were those who had bought their farms on mortgage during the wars and were now unable to pay the interest. Tenants—particularly those with small farms—were also in difficulties. The Grantham newspaper was full of assignments for debt and at Beckingham in west Kesteven, wrote one correspondent to the Board of Agriculture, 'many respectable tenants are in goal [sic]'. On the Heath even farmers who were, as another correspondent wrote, 'staid and frugal in their habits' were in difficulties. Land which had lately been reclaimed was allowed to go back to gorse or warren, and a number of landlords were unable to let their farms. There were similar difficulties in the fenland. Lord Castlereagh, who had a substantial estate near Holbeach, was quite unable to find tenants and one reporter noticed eight farms all unlet within a few miles of Boston. Things were as bad in the Kesteven fenland. Tenant farmers in Branston 'were all in serious distress', at Horbling 'they were all considerable losers' whilst Earl Fortescue's tenants in Billingborough were in a 'precarious and delapidated condition'.[1]

It was generally agreed that the distress was greater on arable than grass farms, but nevertheless there were few parts of Lincolnshire untouched by the fall in prices in 1815 and 1816. Part of the difficulties of the area stemmed from the troubles of the country banks. During the wars these banks had given credit to farms 'to an unwholesome degree'; when they closed on debtors after 1815 they left many farmers bankrupt. Further the banks themselves were in trouble, and a number in South Lincolnshire failed. The collapse of three banks in Boston had particularly unfortunate consequences in the district. In the country as a whole there was a shortage of bank notes, and this prevented farmers getting a fair

---

[1] Board of Agriculture, *The Agricultural State of the Kingdom in 1816* (London, 1816), pp. 150–67; White; L.A.O. Cragg 1/8.

The Agricultural Revolution in South Lincolnshire

price for their products. The diminution of a circulating medium was particularly serious in Lincolnshire where the shortage of notes amounted to two and a half million sterling. Although the country banks recovered later, they were not surprisingly wary of lending to farmers, and credit was short until the establishment of joint stock banks in the 1830's.[1]

By 1817 a brief recovery in wheat prices alleviated conditions. There were, for example, few farms left vacant in the fens that year. But this grace was short-lived, for wheat prices plunged to a remarkably low point in 1822. Years later, when the bad times had gone, 1822 was remembered in Lincolnshire as the most disastrous year in living memory. Many tenants accumulated rental arrears which were to dog them for years. The intensity of the depression varied from district to district. Around Sleaford, for example, conditions were not too bad. The local landlords had responded to low prices by reducing rents, whilst most of the area was tithe free and the poor rates low.[2]

In the later 1820's grain farmers were comparatively free from distress; it was a period of stagnancy. But for the first time graziers were in serious trouble as wool prices fell, and this was made worse by the devastating sheep rot of 1827. Two South Lincolnshire graziers appeared before the Select Committee on the state of the British Wool Trade in 1828. One, John Calcraft of Ancaster, had given up farming in 1826 because he 'was losing money every day'. The other, Richard Creasey, who farmed in south-east Kesteven, was also in great difficulties; he kept his farm on solely because he had ten sons! The poor prices for wool encouraged many farmers to plough up their grassland in order to grow wheat. To do this they generally had to have their landlord's permission. A tenant of Lord Willoughby's wrote a particularly pathetic letter to the estate agent on this subject: 'And the Grazing His so bad I hope your Lord-ship will give me leave to plow sum moor before I ham quiet rewened I have Hall my bread to bye for nine in Famely.'[3]

If the tenant—who wrote in 1831—did in fact eventually plough

[1] Pressnell, pp. 345–9; Board of Agriculture, *Agricultural State*, p. 6; Porter, *The Lincolnshire Magazine*, III; *ibid.* V; B.P.P. III (1836), part I, Q. 7841–6, 80, 85–7; L.A.O. T.L.E. 38/17.   [2] B.P.P. V (1818), 203; Creasey, p. 369.
[3] B.P.P. III (1828), 17–30, V (1833), Q. 12, 212; L.A.O. 2 Anc 7/4/1/.

his grassland, then he suffered from the fickleness of prices at this time for in the following year wheat began to fall as wool began to rise. The mid-1830's were generally bad years; fen farmers in particular seem to have suffered. By 1836 one Spalding merchant reckoned fen farmers were a third worse off than in 1818, and many farmers were paying their rent from capital. Only the large graziers were paying their way and many small tenants had gone bankrupt. The later 1830's brought a respite; but in the early 1840's the uncertainty about the repeal of the Corn Laws brought trouble to many farmers. In fact prices did not fall immediately after repeal, but in 1848 there was again a collapse in wheat prices. Farming in South Lincolnshire seems to have suffered all the symptoms of depression; farms were left untenanted, arrears accumulated and the land was poorly cultivated. But with the onset of the Crimean Wars wheat prices rose again, talk of distress disappeared, and until the late 1870's farmers seem to have enjoyed considerable prosperity.[1]

Although it cannot be denied that there was some agricultural distress in South Lincolnshire between 1815 and 1852, it should be remembered that the witnesses before the Select Committees were more concerned in gaining some government assistance than giving a true picture of conditions. Fortunately the course of the depression can be traced more accurately, for the rental accounts of three great South Lincolnshire estates—those of the Duke of Ancaster, Sir William Welby and Sir John Thorold—are available. They cover respectively the years 1814–97, 1817–87 and 1832–1913. Two graphs have been compiled from these accounts.[2] The figures refer to the acreage of the estate at the beginning of the series and later additions have been excluded. Fig. 16 shows the rental receipts for each year. Fig. 17 shows the accumulated rental arrears owed to each landlord in each year. Both graphs allow a little more precision to be given to the nature and extent of the depression.

Only the Ancaster records contain information on the period before the final break in grain prices at the end of the war, so that

[1] *B.P.P.* III (1836), part 1, Q. 7790; L.A.O. 2 Anc 7/10/45; T.L.E. 38/7, 2 Anc 7/19/21; Thirsk, p. 306.

[2] L.A.O. 3 Anc 6/242–307, 2 Thorold 18–70, Lind Dep. 24/Accounts; *B.P.P.* XVI (1894), part 1, pp. 435–6. For a discussion of these accounts, see D. B. Grigg, 'A Note on Rent in Nineteenth century England', *Agricultural History*, 39 (Davis, California, July, 1965).

The Agricultural Revolution in South Lincolnshire

it is impossible to compare the receipts during the depression with this presumably high level. However, a similar series of rental receipts for the nineteenth century—those compiled by R. J. Thompson in 1907—run from 1801 to 1900, and shows that rents did not fall off until after 1820.[1] If we make a similar assumption for the

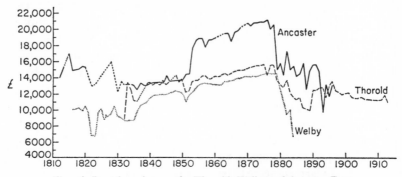

Fig. 16. Rental receipts on the Thorold, Welby and Ancaster Estates.

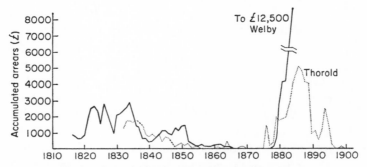

Fig. 17. Tenant debt: the Thorold and Welby estates.

estates discussed here, then the average receipts for the years 1817–20 can be taken to be representative of the level of rents in South Lincolnshire during the later part of the wars. It was not, in fact, until 1822 that rents fell substantially on either the Welby or the Ancaster estates. In that year the Ancaster receipts were 85 per cent of the war level, the Welby receipts but 60 per cent. In the later 1820's the receipts on both estates recovered but were still below the war level. In the early 1830's there was again a sharp decline on the

[1] R. J. Thompson, p. 590.

Paradox and Progress

Ancaster estate even below the low point of 1822. Recovery occurred in the later 1830's, but thereafter the course of rental receipts on the two estates differed. The Ancaster receipts increased very slowly and it was not until 1851 that they exceeded the wartime level, having been below this for over 35 years. On the Welby estate, however, there was a rapid rise from 1836 until 1879, with only a temporary setback in 1851. The Thorold records only begin in 1832, and in 1837 part of the 1836 receipts were received. This accounts for the violent fluctuations in those two years. However, from 1837 onwards the trend was steadily upwards with only a slight fall in 1851. There is unfortunately no means of knowing how these receipts compare with those before 1832.

Rental receipts—the actual amount a landlord received from his tenants in any one year as distinct from the contractual rent—make no allowance for reductions in rent or abatements, so they do not fully measure a landlord's losses. Nor are they an altogether reliable index to the condition of the tenants on an estate. Fig. 17 gives a different picture; all the arrears owed to a landlord in any one year are plotted. The graph shows that whilst tenant debt was already growing before 1820, it did not become considerable until after that year. Debt continued at a fairly high level throughout the 1820's and 1830's, and indeed only disappeared finally after 1850.

The two graphs suggest that whilst there was some distress from 1815 right through until the beginning of the 1850's, the worst times were in the early 1820's and 1830's. The debts accumulated by tenants in these periods were often carried throughout much of the period. There was also certainly a decline from the prosperity of the wars in which both tenant and landlords had shared. Receipts on the Ancaster estate were below the war level until 1851 and on the Welby estate until 1837. But only in a few years did rents fall catastrophically below the war level.

'You Lincolners', wrote a Governor of Christ's Hospital to the agent of their Skellingthorpe estate in 1844, 'certainly appear to have carried on your farming operations with great spirit and deserve to receive great credit for your improvements.'[1] As will be shown

[1] L.A.O. T.L.E. 38/7.

The Agricultural Revolution in South Lincolnshire

in the next chapter, there was a rapid and widespread improvement in Lincolnshire farming between 1815 and the 1840's. At the middle of the century most Lincolnshire men thought the progress made since 1815 was greater than that in the preceding years, and there is little doubt that they were right. Yet this was achieved in times which were apparently far from prosperous for either tenant or landlord. There appear to be two types of explanation for this. In the first place not every region suffered equally from the low prices; this will be discussed in chapter x. In the second place, there were a number of economic characteristics of the period which affected the whole area, and help explain the paradox of poor times but good farming.

In the first place it should be remembered that the depression period of the 1820's coincided with an important point in the extension of the agricultural area of South Lincolnshire. Enclosure had been virtually completed by the end of the wars, and although some heath land remained untouched and additions were made later by embanking the coastal marshes of the Wash, the maximum agricultural area had virtually been attained by 1825. In the preceding years the simplest way of increasing total output had been to plough marginal land; as this was no longer possible, farmers had to seek other ways if output was to continue to increase. There were two obvious alternatives. First, the system of farming could be altered by reducing the grass acreage and growing more fodder crops. This 'mixed' farming would produce more livestock from a given area than where the land was under grass alone; in addition the farmer had the grain crops grown in rotation as a cash crop. This alternative farmers did in fact follow, and this change will be discussed later.

In the second place, the output of the existing acreage could be greatly increased by using more fertilizer, by underdraining and by cultivating the land with greater care. This alternative was also followed. The combination of altering the land-use system and increasing yield per acre resulted in the High Farming of Caird's England, and in Lincolnshire it reached one of its peaks in the 1850's and 1860's.

But even if the maximum agricultural area had not been reached in the 1820's, it is still likely that there would have been an increasing output from agriculture. Demand was still rising rapidly as the

Paradox and Progress

population of England expanded. More important, the farming methods which were being adopted in the 1820's, 1830's and 1840's were not on the whole new. They had been known to some farmers for many years before 1815 and practised for some time by a few. The momentum of change had in fact set in long before the fall of prices, and as Sir John Clapham has written, '...low or fluctuating prices for the staples of agriculture were not sufficient to stop movements already under way'.[1] In some parts of South Lincolnshire good farming was almost forced on farmers. Towards the end of the war a great deal of land was ploughed up on the Heath and sown to endless corn crops. The resulting decline in soil fertility and crop yields convinced many farmers that these thin soils could only be profitably farmed if the rules of good husbandry were followed. Later in the century Samuel Sydney, referring to the Wolds and the Heath, put the matter succinctly: 'In some districts, especially where a light soil gives no choice between large farms, much manure and numerous stock or no cultivation at all...good farming is the rule'.[2]

That most South Lincolnshire farmers chose good farming is indisputable. Nevertheless, the logic of increasing expenditure to increase output per acre seems questionable in the light of the reduced profit margins farmers appear to have been experiencing after 1815. There seem to be two possible explanations.

When farm prices fall and farmers seek to economize, the most obvious way is to reduce labour costs. This was, however, difficult for South Lincolnshire farmers. Probably at least half the farmers in the area relied on family labour alone. Further the larger farmers were hampered by the labour rate or 'roundsmen' system, the method of Poor Law Relief prevalent everywhere except in South Holland. Under this system the unemployed in a parish were allocated to each farmer in proportion to the rateable value of his farm. Although farmers thereby obtained cheap labour, for they paid these men little, they were also prevented from cutting labour costs. Redundant labourers simply returned under the aegis of the labour rate.[3]

---

[1] Clapham, I, 455.
[2] S. Sidney, *Agriculture and Railways in North Lincolnshire* (London, 1848), p. 4.
[3] Hasbach, p. 180; *B.P.P.* XIX (1825), 393; XXVII (1850), 138; L.A.O. 2 B.N.L., 2 Anc 7/3/1.

However, the farmer could reduce production costs per bushel by increasing output per acre. The methods by which this was obtained, such as 'claying', underdraining, the careful cultivation of root crops and the use of fertilizers, all used labour prodigally; the roundsmen system at any rate helped reduce this cost. These improvements initially increased expenditure, but in the long run yields increased proportionally more than the increases in expenditure, and cost per bushel fell. This is well illustrated by the effect of 'claying' peat soils in the fens. A farmer who clayed his peat at Digby at the cost of £2 an acre found his yields increased 50 per cent. As the effect of claying lasted between 5 and 10 years, it was clearly worthwhile.

Whether the policy of increasing expenditure to obtain a reduction in cost per bushel was a wise one is arguable in retrospect; but certainly farmers *at the time* believed that it was the only way to meet lower prices. As early as 1825 John Creasey noticed that farming in the Sleaford district had recovered from the 1822 crash because of a 'spirited and well informed tenantry who contend with adverse times...some of them at an expense that is scarcely justifiable with prudence'. Ten years later George Calthorp noticed that some land in the fens only remained in cultivation at the current low prices because improvements had reduced production costs. In the 1840's the fear of Free Trade led to a similar attitude; falling prices could only be met by improvements. An applicant for a vacant farm on the Ancaster estate rather smugly wrote '...I think it requires a tenant to make great improvements in it, particularly under this free trade'. Improvement was recognized to be necessary even on the poorest land. A Doddington farmer wrote in 1835 of a heavy clay farm that '...it was ...so little in the least useful, that any tenant who occupies it, unless he has capital, must cut to lose in times when the produce makes so little money'.[1]

Low prices then seem to have acted as a stimulus to improvement; and at the end of the period of low prices at least two distinguished writers realized this. 'During a period of high prices', wrote James Caird in 1852, 'moderate rents could be paid without the investment of much capital by the tenant. But low prices and universal

---

[1] Creasey, p. 369; *B.P.P.* III (1836), Q. 8088–91; L.A.O. 2 Anc 7/19/6, Jarvis 8; J. Caird, *English Agriculture in 1850–51* (London, 1852), p. 369.

Paradox and Progress

competition compel agricultural improvement. We must either farm as well as our neighbours or be undersold by them.' G. R. Porter, in his book *The Progress of the Nation*, was more explicit:

> ...it is rather to the stimulus of low prices that we must look to provide that increased quantity which is to make up, safely and satisfactorily to the producer for falling markets.... It may be added that the great agricultural improvements, which have taken place since the peace, and which are still in progress, while they negative the notion of an uninterrupted series of losses on the part of the cultivators, are in a great degree, the consequence of the stimulus to exertion supplied by low prices. Had prices continued high, the farmers would, perhaps, have gone on their old course; but with so considerable a fall as they have experienced in the value of their produce, such a course would have been attended with certain ruin, and in this way the improvements they have made may be said to have been forced upon them.[1]

Low prices seem to have acted as a stimulus to agricultural improvement at other times. For example, between 1730 and 1750 grain prices were low, yet investment on the Duke of Kingston's estate in Lincolnshire and adjacent counties rose above the level which obtained in the more prosperous years before and after; this, in spite of a considerable burden of rental arrears which the Duke carried through the depression. It may have been the need to attract tenants which compelled this improvement. A similar phenomenon occurred in Lincolnshire in the late 1870's and early 1880's, for a number of landlords who appeared before the Richmond Commission claimed they had been compelled to increase expenditure in order to attract tenants. Fifty years later East Anglia had just experienced a decade of depressed agriculture, yet the authors of a survey of agriculture there found little evidence of a halt in improvement. 'In times of prosperity', they wrote in 1932, 'agricultural prosperity alters slowly, but as the margin between cost and price narrows, new methods are adopted with a view to maintaining profits.'[2]

Even if it is accepted that falling prices could stimulate improvement, it still seems doubtful that in a period of long sustained decline

---

[1] Caird, p. 504; G. R. Porter, *The Progress of the Nation* (London, 1847), pp. 141–3.
[2] Mingay, *Economic History Review*, VIII; *B.P.P.* XVI (1895), 14–17; *An Economic Survey of Agriculture in the Eastern Counties* (Cambridge, 1932), p. 8.

The Agricultural Revolution in South Lincolnshire

in landlord and tenant income investment should have been increased or even maintained. Thus whilst landlords in Lincolnshire claimed they had increased investment in the depression of the 1880's, the statistics of landlord expenditure available show quite clearly that investment soon followed prices downwards (Fig. 18 and Table 16). Nor for that matter did landlords need to spend money to attract tenants between 1815 and 1852. Except for the period immediately after the end of the wars there was no shortage of applicants for vacant farms; even in the poorest years Lord Willoughby received several applicants for any vacant farm.

Fig. 18. Annual expenditure on the Thorold and Welby estates.

However, the comparison between the two depression periods is to a considerable extent a false one (Fig. 16, p. 124). Although rents certainly fell below the wartime level it was only in 1822 and the early part of the 1830's that they fell substantially. For the greater part of the depression period rents were little more than 15 per cent below the war level. At the same time the general price level had fallen and so some producer goods that landlords bought had also fallen in price. After 1880 the fall in rents was much more sharp, and had not returned to the level of the 1870's even by 1913.

Whilst the evidence on estate expenditure is more limited than that on rents, the graph of expenditure on three South Lincolnshire estates suggests that whilst it may have declined in the early 1820's, it recovered fairly quickly and from 1830 the trend was steadily upwards. The expenditure in Fig. 18 includes both maintenance costs and improvements; in Table 16 the cost of improvement alone on an estate near Grantham is shown. Whilst no information is available for the period before 1830, it shows that the outlay per annum was quite as high in the depressed 1830's as in the better years between 1842 and 1874; thereafter expenditure fell off rapidly.

There was in fact a great difference between the two depression periods; it would be a mistake to argue that improvement must have halted between 1815 and 1852 because improvement was slowed down in the second depression period. F. M. L. Thompson has aptly compared the two periods. 'The first depression was selective, moderate, accompanied by continuous expansion of output in many areas, and hopes for recovery were always reasonable; the later depression was general, severe and even catastrophic in its effects; accompanied by continuous contraction of cultivation, and any reasonable grounds for expecting a recovery or even a halt in the decline, were hard to discover.'[1] We can reasonably conclude then, that whilst rents and investment may have fallen below the wartime level, it was only in a few years that rent at any rate was more than 15 per cent below the average of 1817–20.

TABLE 16. *Expenditure on C. Turnor's estate, 1830–93*

	Total expenditure £	Average expenditure per annum £
1830–42	38,000	3,166
1842–74	85,000	2,656
1874–86	21,000	1,750
1886–93	11,000	1,550

Source: *B.P.P.* XVI (1894), part 2, Q. 14, 291–3.

Was this level of expenditure sufficient to account for the remarkable increases in productivity which occurred between 1815 and 1852? There is at least one good reason to believe that it was. Before 1815 both landlord and tenant investment had been aimed

[1] F. M. L. Thompson, *Oxford Economic Papers*, p. 305.

The Agricultural Revolution in South Lincolnshire

at increasing the agricultural area; the major forms of investment were in enclosure and in fen drainage. Both were expensive and both involved considerable outlays in large single sums. After 1815 expenditure was quite different. Enclosure was completed, and so was the most expensive part of fen drainage, the cutting of the drains. The expenditure was now in relatively small but frequent 'lumps'—in underdraining, in 'claying', in the purchase of fertilizers and the erection of steam engines to pump water from the drains. Thus to obtain the great rise in productivity of the period there was no necessity to have an increase in investment; it was caused simply by a change in the *direction* of investment.

One last motive for landlord investment may be briefly touched upon. Many landlords in early nineteenth-century England lived in perpetual debt on their overall accounts (that is farm plus non-farm sources of income) and lived on a permanent system of deficit financing. In the 1820's many were particularly acutely confronted with this problem, and were forced either to sell their land or to improve its capital value. It has been suggested that this may partly help explain the great surge of improvement on some large estates at this time. This form of enforced landlord improvement had its counterpart in the early eighteenth century. Professor Habbakuk has observed that at this time many large estates were burdened with debt because of dowries, settlements and portions, so landlords tried to raise the capital value of their estates. There is no evidence as to whether the great landlords of South Lincolnshire were in such circumstances as to prompt improvement, but it may have been a minor factor in accounting for the continued improvement during this period.[1]

So far landlord investment alone has been considered. Before 1815 tenants were rarely prepared to make any considerable outlay, for they had neither long leases nor any alternative form of security for the expenditure they made. Yet after 1815 tenant investment grew considerably. There is no doubt that one reason why this occurred was the establishment in South Lincolnshire between 1810 and 1825 of a system of tenant right.

Tenant right simply meant a recognition that a tenant was en-

[1] Thompson, *Oxford Economic Papers*, p. 36; Habbakuk, p. 15.

Paradox and Progress

titled to compensation for any unexhausted improvements he had made when he left a farm.[1] The practice grew up in areas where there were no long leases, which provided the security a tenant needed to make substantial outlays. In eighteenth-century Lincolnshire there were few long leases and no tenant right. On the other hand tenants often made covenants with their landlords to undertake specific improvements. Otherwise they relied on the good faith of their landlords. But not unnaturally tenants tended to make only those improvements which brought them immediate benefit and this generally meant reclaiming marginal land. Long-term improvements were either left to the landlord or neglected altogether.

Tenant right probably grew out of the right of entry. If a tenant's agreement was ended, he left the farm in April. Thus the incoming tenant had either to be allowed on the land before the outgoing tenant quitted, or the latter had to undertake labour from which the incoming tenant would derive the benefit. This led to a system of compensation between tenants. When the system of tenant right developed in South Lincolnshire the payment of compensation was typically between tenants, not between landlord and tenant. When a tenant of Lord Willoughby claimed compensation from him, Lord Willoughby's agent was most surprised. '...I know of no precedence', he wrote, 'for this claim on the landlord.' But there were cases of landlords paying the compensation. The Governors of Christ's Hospital estate paid one of their tenants for a building he had undertaken. But they immediately charged the incoming tenant the same amount. Although most of the Lincolnshire witnesses before the Select Committee on Agricultural Customs were certain the general practice was for compensation to be between incoming and outgoing tenants one or two witnesses noticed exceptions. For example, the Rev. C. Neville opposed making tenant right a legal contract on the grounds that a landlord might find himself saddled with compensation for several tenants simultaneously.[2]

Tenant right and the great improvements in South Lincolnshire evolved at the same time. But contemporaries agreed that the

[1] On this subject see D. B. Grigg, 'The development of tenant right in South Lincolnshire', *The Lincolnshire Historian*, II, no. 9 (1962), 41–7.

[2] C. S. Orwin, 'The history of tenant right', *Agricultural Progress*, XV (1938), p. 147; *B.P.P.* VII (1847–8), Q. 7275–304.

The Agricultural Revolution in South Lincolnshire

'custom of the county' had not caused the improvements, although its existence must surely have encouraged them. By the 1840's the custom was well established, although it had no legal basis and remained simply an accepted understanding between landlords and tenants throughout the county. The first instances of the system appear to date from between 1815 and 1825. A tenancy agreement between Colonel Jarvis and a tenant called John Nesbitt provide an excellent early example. In 1825 Nesbitt agreed to take a farm on Doddington Moor. He told Colonel Jarvis's agent that it needed a great deal of capital sunk into it before he would get a return. He finally agreed to take the farm at £250 a year provided that £200 of the first year's rent was returned for improvement. He would then take on the land at an increased rent of £270 a year. But if this was not acceptable to the landlord he would want two disinterested farmers to look the farm over at entry; if he then had to leave the farm before he benefited from the improvements he had made, he could then gain fair compensation. This was the essence of tenant right as it developed in South Lincolnshire.[1]

The normal practice was for both the incoming and outgoing tenants to appoint a valuer who estimated the value of the improvements and agreed on a price to be paid by the incoming tenant to the outgoing tenant. The main forms of compensation, in addition to that paid for the labour done for the incoming tenant were for underdraining and the use of feed and fertilizers—guano, bones, lime, oil-cake and claying. A scale of compensation was worked out over the years. The amount allowed decreased with the age of the improvement.[2]

Not surprisingly some valuers were more concerned with getting the best price for their client than estimating the true value of the improvement, and so the system was inevitably open to abuse. Tenant right was practised in Surrey and Sussex, and according to James Caird was frequently subject to fraudulence. He admitted, somewhat reluctantly, for he was not an advocate of tenant right, that there was little evidence of malpractice in Lincolnshire. Most

[1] Caird, p. 185; *B.P.P.* VII (1847–8), Q. 5968, Q. 7152–4, 7356, 7461–6; L.A.O. Jarvis 8.

[2] *B.P.P.* VII (1847–8), Q. 209–469, 470–681, 7146–239, 7240–306, 7344–441, 7456–553.

Paradox and Progress

of the difficulties arose from the lack of any written agreements or an approved scale of allowances. The scale of payments evolved through a series of disputes and disagreements, and not until 1875 was a written code available. An example of how allowances were decided upon can be traced in a series of letters written between Richard Carline and the Governors of Christ's Hospital. An outgoing tenant on their Skellingthorpe estate had claimed from the landlord for both oil-cake and straw manure. The Governors eventually decided '...it appears to us that the Governors are thus called upon to pay for the same article twice over, once before it goes into the mouths of the cattle and afterwards as manure.'[1]

There were less savoury wrangles over tenant right. Mr Hodson, a tenant of Lord Willoughby tried to claim compensation from the incoming tenant for a building erected on his farm. The incoming tenant's valuer argued that if any part—however small—of the building was financed by the landlord, then it all belonged to the landlord and no compensation was due. It transpired that Hodson had agreed to erect the building at his own expense, on the express condition that no compensation was to be paid. Hodson had not mentioned this to either valuer. He was not an altogether reliable tenant, for a few years later he was again found in a dispute. This time he and the incoming tenant had conspired to defraud the landlord. 'They agreed to try it on, as they say in these parts', wrote Lord Willoughby's agent.[2]

Why should tenant right have developed at this time in South Lincolnshire? Many parts of England had neither leases nor tenant right until the 1880's, and there is no doubt that the existence of tenant right must have been a major factor in accounting for the rapid transformation of Lincolnshire farming between 1815 and 1850. First, there were few leases in South Lincolnshire; so that tenant right was necessary. Secondly, there was a great deal of land in the area which could not be worked without substantial investment by the tenant. In the eighteenth century the tenant had relied on the good faith of the landlord, and many tenancies were handed down from father to son.[3] But after 1815, when prices were poor

[1] Caird, p. 185; L.A.O. T.L.E. 38/7. [2] L.A.O. 2 Anc 7/19/71, 75.
[3] *J.R.A.S.E.* IV (1843), 299.

and there was more danger of eviction, tenants naturally wanted some form of protection.

It is possible now to return to the apparent paradox remarked upon at the beginning of this chapter. Why did farming productivity increase so rapidly in a period of depression? It is not suggested that any of the explanations offered is definitive, but certainly considered together they help account for the behaviour of farmers at this time.

CHAPTER VIII

THE AGE OF IMPROVEMENT

'If Norfolk has long held the first rank among the English counties for agricultural development, Lincolnshire, which a century ago was more waste and sterile, now disputes the palm.' So wrote a French visitor to England in the 1850's. James Caird was more sceptical and thought Lincolnshire farming was more characterized by the rapidity of its improvements since 1815 than by any particular excellence it had in the middle of the century.[1] But there seems little doubt that by the 1850's and 1860's the county was one of the best farmed regions in England, and certainly the contrast with the beginning of the century was prodigious. Perhaps there had been few striking innovations; but the new methods of the early part of the century had been generally adopted, and it was to this uniformly high standard rather than to the efforts of a few great innovators that South Lincolnshire owed its reputation.

Nowhere was this more true than in the fenland, for the great fen engineers were associated with the schemes of the late eighteenth century and early nineteenth century; these were completed by 1807, and the network of drains achieved by then remains much the same today. In the 1830's and 1840's fen engineers were primarily concerned with making the existing system work. There were three ways in which they attempted this; first, the outfalls of the main rivers were deepened and straightened; secondly, the main drains and rivers were scoured out at more regular intervals; and third, the inefficient windmills were replaced by steam engines.

The importance of the outfalls had been recognized at an early date. George Maxwell wrote in 1791; '...let the rivers and outfalls be attended to and the banks will in great measure take care of themselves, for let it be remembered that every inch taken out of a river is an inch added to the strongest part...and particularly let

---

[1] L. de Lavergne, *The Rural Economy of England, Scotland and Ireland* (Edinburgh, 1855), p. 231; Caird, p. 193.

it be remembered after all that can be done in raising and strengthening of banks, the work is a mere palliative, for when our money is wasted, it is on the outfall and the outfall only that we must at last depend for our security'.[1] By far the most important single outfall was that of the Witham, Boston Haven, for the Witham Fens, East, West and Wildmore Fens and the Black Sluice Level all drained into it. The importance of improving the Haven had been recognized at an early date, yet, due largely to the division of control it was not until the 1880's that the Haven was properly improved. Fortunately a great deal of piecemeal work went on during the early part of the nineteenth century and this was enough to improve the drainage of the Interior Fens. In 1825 a new cut was made for the Witham from the Grand Sluice to the Ironbridge. Between 1827 and 1833 a more substantial advance was made when the great bend at Wyberton Roads was cut off by the construction of a new channel. In 1841 a further cut was made and the channel carried as far as Hobhole Sluice by fascine training. All these improvements shortened the lower outfall and benefited the drainage of the fens; but it left the circuitous course from Hobhole to the outfall proper at Clayhole untouched;[2] and it was not until 1884 that this problem was attacked.

The Witham itself was not neglected. The works which had been suspended in 1816 (above, pp. 27–8) were renewed in 1825. The whole course of the river from the Grand Sluice in Boston to Lincoln was deepened and embanked. A new channel was constructed from Bardney to Lincoln in order to separate the navigation from the drainage waters. These works, of course, improved the drainage of the Kesteven and Holland fens to the south of the river, but had no effect on the East, West and Wildmore Fens which drained directly into the Haven. Ironically these fens whose initial drainage had been so successful, declined after 1807. This was due partly to the slow silting of the Haven, which blocked Hobhole Sluice; but this was ameliorated by the work on the Haven described above. More serious, although less obvious in its impact, was that the success of the early drainage caused the peat in the East Fen to shrink or settle; thus the gradient between East Fen and the Haven

[1] S.G.S./S.R./5/37. [2] Wheeler, pp. 348, 352–5.

was slowly reduced. This problem was not really solved until the belated introduction of steam pumping in 1867.[1]

Steam pumping had come earlier elsewhere. The first steam pump in the Witham Fens was erected in 1832, and by 1851 there were eleven between Lincoln and South Kyme. Water in the parish drains could now be moved into the Witham much more quickly and without any reliance on wind; there was no question of work being held up in springtime because the flood waters could not be lifted. But these parishes still depended on the ability of the Witham to move this water out through the Grand Sluice, and so silting below the sluice could still cause the Witham waters to back up and overflow.[2]

The Black Sluice Level between Boston and Bourne benefited greatly from the improvements to Boston Haven, but the Level still suffered from flooding. The South Forty Foot Drain was over 21 miles in length, and the southern end was neglected by the Commissioners. Thus two plans to scour and embank the whole length of the drain in 1831 and 1838 were only carried out for the section near Boston. The troubles of the southerly section were increased by the risks of flooding from the river Glen. Twice in the 1830's the river breached its banks and overflowed the Level. The whole Level suffered because land to the east of the Hammond Beck which formerly drained directly into the Wash now drained westwards into the South Forty Foot. Not only did this extra water overload the capacity of the drain, but the Commissioners received no taxes from the occupants, for the Hammond Beck still marked the eastern boundary of the Black Sluice Level.[3]

Until 1841 there were no steam pumps anywhere on the Level and the waters of the parish drains were lifted into the South Forty Foot by 63 windmills. In that year the Trustees of the Bourne drainage district decided on their own initiative to erect a steam pump. The Level Commissioners, fearing this would overburden the drain, opposed the measure; a great deal of litigation followed, demonstrating once again the lack of co-ordinated control of the drainage system. A steam pump was eventually erected, but the dispute prompted the Commissioners to put their own house in

---

[1] Wheeler, pp. 166, 231–7.   [2] *Ibid.* pp. 186–91; *J.R.A.S.E.* XII (1851), 328.
[3] L.A.O. Whitfield XV/2.

order. Three reports were made on the Level; they all concluded that the main cause of flooding was the overriding of the fen waters by highland waters. The obvious cure for this was to scour out the old Car Dyke, but the taxpayers considered this too expensive. Finally, under an act of 1846 the whole of the South Forty Foot was deepened and a new sluice constructed south of the old Black Sluice. This improved the drainage but it was still not natural; forty windmills had to pump water into the Drain. By 1851 they were being replaced by steam engines.[1]

The drainage of Deeping Fen was still ultimately dependent on the outfall of the Welland, and the failure to carry out the proposed deepening and embanking began to have serious effects. In 1837 a scheme to train the outfall channel from Fosdyke Bridge to the Witham was begun. Like so many fen schemes it was never completed, but the new channel—2 miles of fascine training—certainly lowered the bed of the Welland. This allowed both Vernatts Drain and the River Glen to empty more easily into the Welland, with beneficial results to Deeping Fen. The Trustees of the Fen had attempted to improve their drainage before the deepening of the Welland. The great scheme of 1801 had been less successful than hoped, largely due to the inadequate fall in gradient between Podehole and Vernatts Sluice. This meant that the Welland waters constantly held back the Deeping waters in Vernatts Drain. In 1824 it was decided to raise the head in Vernatts by erecting steam pumps at Podehole. At first this was not successful, but after the North and South Drove Drains had been scoured in 1830 the drainage did substantially improve.[2]

But the most radical improvement in fen drainage came in South Holland. The cutting of the South Holland Main Drain had not led

[1] L.A.O. Whitfield XV/1/20 and 37; Wheeler, pp. 265–9; W. Lewin, *Report to the Commissioners of the Black Sluice Drainage* (Boston, 1843); Sir John Rennie, *Report on the Black Sluice Level* (London, 1845); W. Cubitt, *Report on the Black Sluice Drainage* (Boston, 1846).

[2] J. Walker, *The Welland Navigation* (London, 1838); idem, *Report to the Trustees of the Outfall of the River Welland* (London, 1835); Wheeler, pp. 302–5; T. Pear, Jr., *Report on the Improvement of the Nene Outfall and North Level Drainage* (Bourne, 1845); idem, *Report on the Improvement of the Drainage of Deeping Fen by Steam Power* (Bourne, 1823); J. Glyn, 'Drainage by steam power', *Transactions of the Royal Society of Arts*, LI (London, 1837); W. S. Mylne, *On the Steam Engine Drainage of Deeping Fen* (London, 1830).

The Age of Improvement

to the improvement which was hoped for, largely due to the poor state of its outfall, the river Nene. In 1831 a new outfall for the Nene was cut from Gunthorpe Sluice right out across Cross Keys Wash. This lowered the level of low water in the channel by 10 feet and allowed the sluice of the Main Drain to be opened whenever necessary, greatly improving the drainage of South Holland.[1]

The result of all these schemes was greatly to improve the drainage of the Interior Fens. The deepening of the outfalls meant that the main drains could be emptied more frequently and there was less risk of waters backing up and overflowing the lower levels. The construction of steam pumps guaranteed that parish drains could be emptied into the main drains, and so water was not left standing on the fields in spring. The frequency of flooding was greatly reduced in this period, and although the final improvements were not made to the fen drainage system until 1884, the risk of flooding had ceased to be the prime concern of fen farmers by mid century. The greater security had, as will be seen in chapter IX, a profound effect on land use.

Whilst fen farmers were preoccupied with the surface drainage of their land, farmers in much of the rest of South Lincolnshire were more concerned with 'hollow' or underdrainage. The limestone, marlstone and gravel soils of Kesteven generally drained naturally—indeed in hot summers too freely—but on the clays of west and south-east Kesteven and also on the heavier silts of the fenland, underdrainage was the prime prerequisite for good farming. The history of underdrainage is so much associated with the introduction of the pipe, and the government loans of the 1840's, that it is often thought of as beginning at that time, but in Lincolnshire underdrainage probably began in the 1820's. Certainly there was little drainage at the beginning of the century when the common form of drainage was ridge and furrow which, of course, only affected surface water. There is some evidence of hollow drainage in the 1790's, but only among the more enterprising farmers.[2]

[1] Wheeler, pp. 112–13.
[2] On underdrainage in general see B. Adkin, *Land Drainage* (London, 1933); G. E. Fussell, *The Farmer's Tools* (London, 1952), pp. 15 ff.; H. Nicholson, *The Principles of Field Drainage* (Cambridge, 1942).

The Agricultural Revolution in South Lincolnshire

In the 1820's underdrainage was to some extent forced on clay farmers, for many of the new farming methods were largely impracticable on undrained clays, whilst the fall in prices encouraged any methods which increased crop yields. In the 1820's two particularly disastrous attacks of sheep rot prompted a wave of underdrainage, and this carried on through the 1830's and 1840's. When exactly underdrainage began on a large scale it is difficult to say, but certainly on the Jarvis estate at Doddington underdraining of clay farms was common from 1812 onwards. Sir John Thorold, who had several farms in west Kesteven, was encouraging his tenants to underdrain at least as early as 1832; south-east of Sleaford the small outcrop of tenacious and wet Oxford clay was first drained in the early 1820's. On the boulder clays of south-east Kesteven there was less done, although Lord Willoughby was certainly prompting his tenants to underdrain through the 1820's and 1830's. In the fens farmers were still primarily concerned with surface drainage until about 1840. But the heavy silts, particularly the blue buttery clays, certainly needed underdraining, and in the 1840's there was a determined campaign to carry this out. When Caird visited East and West Fens in 1851, he noticed that the depression had hit mainly those small farmers who had not drained, whilst those who had done so were doing well. On the peat soils—which were dwindling rapidly by 1851—underdrainage was thought necessary, and certainly the peats did drain fairly freely once the surface drainage was secure. Later however, in the 1860's, the peat soils were underdrained as well.[1]

A number of methods of underdraining were current in nineteenth-century England, and they were all used in South Lincolnshire. The most common and probably the most economic was tile drainage. The first recorded instance in the area is in 1812; by the 1820's all the great landlords were advocating this system. The cost was normally divided between landlord and tenant, the landlord providing the tiles and the tenant laying them. This was extremely expensive in terms of labour as they were all laid by hand. The massive scale of the campaign to underdrain can be measured by the number of tile factories which sprang up. One near Spalding made

[1] L.A.O. Jarvis 8, 2 Anc 7/17/6, 7/7/55, 7/19/57, 7/27/30; *Thorold Estate Survey, 1832*; Creasey, p. 369; Caird, p. 187.

200,000 tiles in 1847; Lord Willoughby had his own tile works on his estate and there were several works on the Lias clay plain of west Kesteven.¹

Another way was to dig a ditch, fill it with stones or thorns and then replace the turf. Less effective was sod drainage; a V-shaped drain was cut, and simply covered by the sod which had been removed to form the drain. Neither of these methods was as effective as tile drainage but they were a great deal less expensive. A Heckington farmer experimented with all three methods on his clays; to drain an acre of clay arable land with tiles cost him £3. 2s. 6d.; with thorns, £1. 17s. 6d. and with sods 18s. 6d. Furthermore, the sod drainage included the cost of the main drain whilst the other two did not.²

Some form of pipe drainage had been used in England since the eighteenth century, but the high cost prevented its general adoption until Scraggs' pipe-making machine was developed. Equally rare was mole-ploughing. One Aslackby farmer bore-ploughed his grass land as early as 1831 but he must have been an exception; on the other hand, a great deal of Kirton Hundred was being mole-ploughed in the 1840's so that this practice may have been going on elsewhere.³

By the middle of the century some underdraining had been carried out in all the clay regions, but with varying degrees of completeness. The most successful draining was in west Kesteven. The whole area had been underdrained by 1851, but with tiles in shallow drains. In the 1840's opinion was veering to Parkes' view that drains should be deep, and landlords were beginning to re-drain. This was probably quite unnecessary as modern drains are approximately the same depth as those in the first drainage of the area. Landlords in west Kesteven were particularly active in encouraging underdrainage, and Lord Brownlow's steward, Samuel Hutchinson, wrote two influential pamphlets on the subject.⁴

¹ L.A.O. Jarvis 8, T.L.E. 38/7, 2 Anc 7/27/30; A. E. Trueman, 'The Lias brickyards of west Kesteven', *Transactions of the Lincolnshire Naturalists Union*, IV (1916-18); Anon., *The Effect of Free Trade on the Various Classes of Society* (Spalding, 1849).
² *J.R.A.S.E.* II (1842), 165-8.   ³ *Ibid.* XII (1851), 385; L.A.O. 2 Anc 7/7/6.
⁴ *J.R.A.S.E.* VII (1851), 376-7; Samuel Hutchinson, *Practical Instructions on the Drainage of Land* (Grantham, 1847).

The Agricultural Revolution in South Lincolnshire

In south-east Kesteven less had been achieved. Although some of Lord Willoughby's tenants had drained, the general picture was gloomy, '...the clays are generally without any form of underdrainage', wrote John Clarke of this region. Most of the clays were still in lands. This neglect had deplorable results. After the wet summer of 1853 Lord Willoughby's steward wrote that '...all the undrained districts cut a most miserable appearance this year, and the yield will be much below the average...the cornfields on all cold wet claylands are very full of weeds and rubbish and the turnips generally are in many places all smothered up'.[1]

As was noticed earlier the need for underdraining was realized late in the fenland, but after 1840 it was most rapidly carried out. The clay fens of East and West Fen were being drained in the 1840's; by 1848 Kirton Hundred had been largely underdrained with tiles whilst Deeping Fen had been partially drained by tiles and thorns. In South Holland, however, there had been virtually no attempt to drain, and the landscape presented as gloomy a picture as southeast Kesteven: '...the pastures abound in low places and long hollows, which the rain always fills with water. The arable fen clay is difficult to work, always either miry or wet. A deep subsoil drainage and a deep pulverization, the grand requisites of this district are entirely unattempted on an efficient scale....'[2]

Whilst drainage improvements were essential if any further advances were to be made the most striking improvement between 1815 and 1850 came with the spread of mixed farming. Philip Pusey had the gist of this new style of farming when he wrote in 1842: 'It is well known that the stock which furnished our forefathers with meat were fattened on rich grasslands, but that by a great revolution in farming the light arable soils now chiefly supply the country with animal food'. Mixed farming meant essentially the integration of livestock and crop husbandry on the same farm to their mutual benefit. A greater proportion of the farm was put under the plough, but much of the new arable was devoted to fodder crops; the root break gave

[1] *J.R.A.S.E.* XII (1851), 379–80; L.A.O. 2 Anc 7/27/37.
[2] P. Thompson, p. 694; *J.R.A.S.E.* VIII (1847), 117, 121, 124, 128; XII (1851), 293, 387, 388.

The Age of Improvement

the land a rest from grains and allowed weeding to be carried out, but it also provided a feed for sheep, who in their turn manured the land and helped consolidate the lighter soils. Similarly stall-fed cattle were fed on crops grown on the farm as well as purchased oil-cake; but their dung gave a farmyard manure which helped increase wheat yields. The mixed farming of eastern England at this time was an elegant interlocking system which has been much admired. But with the enormous amounts of labour it required, for hoeing, stone picking, for claying and for underdraining, together with the generous purchases of fertilizers and cattle feeds it may not have been a very efficient system,[1] and indeed proved vulnerable in the depressions of the 1880's, 1890's and 1920's.

At the end of the eighteenth century South Lincolnshire farming consisted of either specialized grazing or rudimentary arable farming where livestock were a rather incidental feature of the husbandry. It is true that some farmers were acquainted with the idea of mixed farming. Nathaniel Kent had pressed the system on a Holbeach landlord in 1778; 20 years later some of the Townland graziers were changing over to mixed farming, whilst there were certainly many farmers on the Heath who had adopted most of the features of the system during the wars.[2] But it was only during the period after 1815 that it became general in South Lincolnshire. The important features of the system which were increasingly practised were: an increasing acreage under roots and seeds; the stall feeding of cattle; the use of artificial fertilizers; and the establishment of improved rotations.

Fodder crops, particularly the turnip, were already grown in South Lincolnshire in 1801 (see above, pp. 76–7) but in the next forty years they occupied an increasing proportion of the arable acreage. Just as important, they were much better cultivated so that the root break served its function as a cleaning crop. Whereas folding of sheep on turnips had been unusual at the beginning of the century, it was the general practice where turnips were grown by the middle of the century.[3]

---

[1] E. L. Jones, 'The changing basis of English Agricultural prosperity, 1853–73', *Agricultural History Review*, x, part ii (1962), 102–19.

[2] L.A.O./T.Y.R./4/1/103; P.R.O./H.O./42/53/99.

[3] *J.R.A.S.E.* XII (1851), 334, 339–40.

The Agricultural Revolution in South Lincolnshire

In the winter cattle were kept in stalls, fed on fodder crops grown on the farm, and, more significantly, on purchased oil-cake. Oil-cake had been fed to stalled cattle on a few Heath farms in the 1800's, but by 1851 it was the general practice in every region except south-east Kesteven. In that sadly backward region many farmers' attitude was summed up by the comments of a tenant of Lord Willoughby's. 'Last year I enquired among the tenants', wrote the estate steward, 'of his Lordships orders as to the probable quantity of oil-cake there might be consumed by each tenant on his occupation; Mr Gilbert then told me he never bought a hundredweight of cake in his life, and did not intend doing so.' But stall feeding on oil-cake was general in the other regions, and indeed Philip Pusey regarded it as the most distinctive feature of Lincolnshire farming; 'this peculiar practice', he called it, for it was only found elsewhere in the Lothians and East Anglia. High Farming was in fact High Feeding. In the Lothians and East Anglia turnips were the main feed for stall-fed cattle, but Lincolnshire farmers thought their light soils too thin to have this crop drawn, and fed sheep off on the crop *in situ*; oil-cake was the main cattle feed. Oil-cake fed cattle gave a fine rich dung which made excellent manure when mixed with straw from the cash grain crop.[1]

The increased feeding of cattle implied, of course, a change in rotations. Even at the beginning of the century some farmers were following the Norfolk four course, particularly on the Heath. But they were a minority. By the 1840's the four course was generally adopted on the Heath. Pusey wrote in 1843 of Lincoln Heath: '...what struck me particularly, you not only see generally very high farming, but you see in forty miles hardly any bad farming—scarcely two or three slovenly fields'. The rotation of (1) turnips, (2) barley, (3) seeds and (4) wheat was to be found nearly everywhere on the limestone soils except around Grantham, where a six-field course persisted; but this was giving way to the four course by 1851. Indeed, everywhere the turnip could be grown the four course was followed; on the Heath and on the gravels of north-west Kesteven, on the gravels of the fen edge and even on the patches of limestone in

---

[1] *J.R.A.S.E.* IV (1843), 330–3; VI (1847), 120, 127, 128, 132; XII (1851), 339, 376, 377, 380, 382, 385, 398–400; Caird, pp. 19–22; L.A.O. 2 Anc 7/9/52.

south-east Kesteven. On all these soils the turnip and the folding of sheep were the central feature of farming. The most significant variation of the four course was the introduction of a second year of seeds, but this did not become general until the depression period at the end of the century.[1]

On the lighter clays—particularly those which were well drained —turnips could be grown, but for the most part the crop was unsuccessful on clay soils. The dead fallow remained an essential part of clay farming in South Lincolnshire; an attempt to grow tares failed. But the backward practices of the 1800's had gone even on the clays. The traditional course of (1) dead fallow, (2) wheat and (3) beans now was only rarely found. A common rotation in west Kesteven was (1) fallow, (2) wheat, (3) seeds or beans and (4) wheat or barley. In south-east Kesteven, on the boulder clays, a typical rotation was (1) fallow, (2) wheat, (3) seeds and (4) oats. The most noticeable feature of these new rotations was the universal year under temporary grasses.[2]

In the fenland there were still a few farmers who regularly pared and burned; and in Skirbeck some farmers still followed three successive grain crops with a fallow. James Caird, in fact, thought fen farming extremely lax; 'The course of husbandry', he wrote in 1851, 'is not very definite, most farmers being permitted to farm as they think best.' Whilst it was true that the fertility of the fen soils did not necessitate the careful farming of the light lands, most fen farmers attempted to follow a rotation. The best fen farming was to be found in Deeping Fen where the course was (1) coleseed, (2) wheat, (3) clover, (4) wheat and (5) oats. Holland Fen and the Black Sluice Level were farmed in a similar manner. The only parts of the fen which really deserved James Caird's strictures were the Witham fens between Billinghay and Lincoln, where there was no regular course. This was largely because fen farms were appendages to farms on the Heath.[3]

By the middle of the century then the majority of South Lincolnshire farmers were following some improved rotation. These were

[1] *J.R.A.S.E.* IV (1843), pp. 288, 303; XII (1851), pp. 339–80.
[2] *Ibid.* XII (1851), 376–80; Caird, p. 181.
[3] *J.R.A.S.E.* X (1849), 119, 123, 126, 128, 132; XII (1851), 383–6; Caird, p. 181.

invariably based on the original Norfolk four course, but because of the relatively small outcrops of light soils, the turnip was not the central feature of the modified rotations. It was the year under 'seeds' which was most characteristic of Lincolnshire rotations. This not only provided grazing and hay, but more important helped replenish the nitrogen content of the soils.

The new rotations alone, together with the application of farm-yard manure, would have raised crop yields substantially; but in addition Lincolnshire farmers were using artificial fertilizers in large quantities in the 1820's and 1830's. Of these the most important was bones. The idea of spreading crushed bones on the land had been brought to the Heath from Yorkshire in the 1790's, but they were not in general use even there until the 1820's. One cause of the slow adoption was cost. At first large lumps of bone were spread on the turnip fallow, but later the bones were crushed and the dust spread. In the 1790's 50–60 bushels were used to every acre; bone dust in the 1800's required 30–36 bushels an acre. The use of the drill to spread the dust was adopted in the 1820's and this led to further economies; only about 20 bushels an acre were necessary. Finally experiments led to the dissolving of the bones in sulphuric acid; between 3 and 6 bushels only were necessary then, and this of course substantially reduced the cost. Bones were most handsomely applied on the Heath, where down to the 1840's they were the main fertilizer used. But they were also used, if less liberally, in most other regions; in west Kesteven, in Deeping Fen, on the gravels of the north-west and in most of Kirton and Skirbeck Hundreds. The most noticeable exceptions were south-east Kesteven, the worst farmed region in South Lincolnshire, the Witham Fens and South Holland. In the latter two regions, according to John Clarke, no artificial fertilizers at all were used.[1]

Bones were the most important but not the only fertilizer used. On the Heath some farmers were experimenting with subsoil ploughing, and in 1851 guano was beginning to be used, although its heyday came later. In west Kesteven lime was used on the grasslands, but in south-east Kesteven where it would have been invaluable,

[1] Creasey, p. 372; J.R.A.S.E. IV (1843), 164; VIII (1847), 120, 123, 127, 128, 132; XII (1851), 339–40, 379–80, 382; Caird, p. 190; L.A.O. 2 Anc 7/9/52.

little was applied, nor was it much used in the fenland. In the fens the most important means of manuring—other than farmyard dung—was the 'claying' of peaty and skirty soils. The original peat soils gave a poor wheat; but when the peats were mixed with clay both better yields and better quality crops were obtained. Claying was not difficult, for the peats were nearly everywhere underlain by blue buttery clay. Trenches were cut through the peat and the clay thrown up into heaps over the field; it was then distributed more evenly over the surface, and ploughed in. Arthur Young had noticed this practice in the Cambridgeshire fens in 1800, but it had not been thought significant. In South Lincolnshire 'claying' seems to have got under way between 1810 and 1825, and by 1851 the whole of the peat and skirty fens had been clayed at least once.[1]

Although claying increased yields substantially, the process was expensive, for the labour needed was considerable. The actual cost depended on the thickness of the peat; where it was thickest claying averaged about £2 an acre; in Dorrington Fen it reached £2. 14s., but in most parts of the fenland £1. 10s. an acre was the standard cost. The results were well worth it. One fenland farmer reckoned that claying had doubled yields, whilst the quality of the crops—particularly of wheat—was greatly improved. The effect of claying was not permanent, and in 1851 many fen farms needed re-claying; some farmers, however, had clayed their peats as often as once every 6 years, and thought the returns well worth while.[2]

'They are an inferior quality of beast, the Lincolnshire beasts; that is all I know about them.' This was the opinion of a butcher who regularly purchased livestock from Smithfield Market in the 1840's. Whilst he may have been uncharitable, there seems little doubt that progress in cattle breeding lagged behind other developments in agriculture in South Lincolnshire; shorthorns were first introduced into the county in the second decade of the century, but it was not until the 1840's that William Torr in North Lincolnshire bred the Lincoln Red Shorthorn. Towards the end of the century the

---

[1] *J.R.A.S.E.* VIII (1847), 120, 128; XII (1851), 341, 376, 380; Caird, p. 183; L.A.O./T.L.E./38/19/17.

[2] *J.R.A.S.E.* II (1841), 406–11; IV (1843), pp. 296–7; VIII (1847), 92–3, 119, 126, 131; XII (1851), 382; Caird, p. 181; *B.P.P.* VII (1847–8), Q. 254, 257, 407–8, 470–1, 428–96.

county's livestock had a much higher reputation. 'It would be difficult to speak too highly of the stock kept on the Lincolnshire farms', wrote an experienced farm judge in 1887. But in the 30 or 40 years after the end of the wars cattle breeds improved but slowly. This was less true of sheep breeding; by the 1820's the old Lincoln Longwool had quite disappeared and the Lincoln–Leicester crossbreed had triumphed. Mutton as well as wool could be got from these animals and they could be fattened much more quickly. In the rest of the century farmers concentrated on developing these characteristics. In the 1870's it could be written that '...the Lincoln takes first place among the longwooled breeds on account of both the weight of its carcase and the quality of its fleece'. In the 1860's and 1870's the export of Longwool rams had become a thriving trade and Lincoln Longwools were sent to start flocks in South America, South Africa, New Zealand and Australia.[1]

It is difficult to trace the adoption of machinery in the area. The most important innovation of the eighteenth century, the drill, was not in general use even in Norfolk in the 1790's; by the 1820's many farmers were using it on Lincoln Heath and the implement spread slowly throughout South Lincolnshire after this. The threshing machine seems to have been first used at the same time; in the second decade of the century there were manufacturers of such machines in Boston and Grantham. Although probably relatively few farmers owned threshing machines, many hired them and in the 1840's the portable threshing machine was a common sight in the fenland. In 1831 there were riots in Kesteven among labourers who thought they were losing their livelihood as the machines spread. The reaper was still largely experimental and it was not until McCormick's reaper was imported from America in the middle of the century that a practical reaper was available. Of the more general implements, it can only be said that progress was at least as rapid as in other parts of England. In 1856 Pyshey Thompson observed that the farmers of Skirbeck, who had been so backward when he wrote his first history of Boston in 1820, were now well provided with such implements as Crosskill's clod crusher, Bentall's

---

[1] B.P.P. XIX (1849), Q. 3, 413; J.R.A.S.E. XII (1851), 393–6, 400; XXXIX (1878), 566, 711; G. Collins, 'The Cattle of the County', *The Lincolnshire Magazine*, I, no. 9 (1934).

The Age of Improvement

broad share plough and the iron harrow. To what extent the small farmers, who were so numerous in west Kesteven and the Fenland, owned implements, it is difficult to say; but the borrowing of implements seems to have been common.

The spread of machinery in South Lincolnshire must have been accelerated by the initiative of a number of enterprising manufacturers in the area. Richard Hornsby founded his firm in Grantham in 1815, Tuxford and Sons were established at Boston by 1842 and in the same year Clayton and Shuttleworth opened a foundry in Lincoln. These and other South Lincolnshire implement makers were regular and successful exhibitors at the Royal Agricultural Society shows in the 1840's and 1850's. Clayton's rapidly became a major concern. In 1850 they sold 140 portable steam engines, in 1855 they sold 491, and between 1856 and 1882, 19,000; between 1849 and 1882 they sold 17,000 threshing sets; not all, of course, were bought in South Lincolnshire.[1]

Whilst steam power had been quickly adapted to fen drainage and threshing, it was less successful in ploughing. Lord Willoughby, however, conducted some of the earliest experiments in England with steam ploughing on his estate at Edenham and exhibited a 'set' at the Great Exhibition in 1851. It was not, however, until after the period under discussion that Fowler's equipment, the most practicable of the many experimental 'sets' devised, came into use in Lincolnshire. The first set was bought in 1859; yet by 1867 there were still only nine sets in use in the whole county; but this was more than in any other county. Only the very large farmers could profitably utilize machinery such as this. And indeed it was only the farmers of 300 acres or more who had much in the way of implements at all even as late as 1881.[2]

There is little doubt that between 1815 and 1851 the farming of South Lincolnshire had been prodigiously improved; by the 1840's

---

[1] Fussell, *The Farmer's Tools*, p. 106; Creasey, p. 370; W. Tritton, 'The origin of the threshing machine', *The Lincolnshire Magazine*, XI, no. 2, p. 19; *J.R.A.S.E.* III (1842), 215; IV (1843), 305; P. Thompson, p. 694; *B.P.P.* LV (1906), Q. 6009; *J.R.A.S.E.* XIX (1858), 316–52; XLIV (1883), 270–6.

[2] G. E. Fussell, 'Steam cultivation in England', *Engineering* (London, 30 July 1943); *J.R.A.S.E.* XXVIII (1867), 198–373; L.A.O. 2 Anc 7/19/45.

The Agricultural Revolution in South Lincolnshire

many informed observers thought that the farming of the Wolds and the Heath had no peer in Britain; and even on the clays and fen farms the standards of farm management had risen strikingly in the space of but 30 years. The consequences of these improvements were considerable. In the next chapter the impact on the agricultural geography of the area will be outlined, but here the effect on farming productivity will be discussed.

In 1848 and 1851 John Clarke published two brilliant essays on the agriculture of Lincolnshire and the fenland; he was careful to discuss the general standard of farming rather than to take only particularly good examples of High Farming. Furthermore, he was a Lincolnshire man—unlike Young or Stone—and had an excellent grasp of the regional differences in the farming of the county. For each region he described he gave the average yield for wheat, barley and oats.

TABLE 17. *Average crop yields, 1848–51 (bushels per acre)*

	Wheat	Barley	Oats
Fen edge (clay soil)	20	—	—
Fen edge (sandy soil)	32	40	56
Marlstone	32	40	52
South-east Kesteven (clay)	26	40	48
Lincoln Heath (limestone)	28–30	40	—
West Kesteven (clay)	28–32	—	—
Fen edge (clay and peat)	32	—	—
East Fen (clay and peat)	36	—	68
Wildmoor Fen (clay and peat)	30	—	50

Sources: J.R.A.S.E. IX (1848), 124, 132; XII (1851), 182, 184, 375, 376, 378, 381.

From the table it seems reasonable to assume that the average yield of wheat in South Lincolnshire in the middle of the century was between 28 and 32 bushels an acre. It will be remembered that the average yield at the end of the war period was about 16 bushels, so that in the space of only 30 years yields had nearly doubled. This, of course, contrasted sharply with the 50 or so years preceding the end of the wars when the average yield rose very slowly (see above, pp. 58–9). There seems no doubt then that the critical

The Age of Improvement

period for Lincolnshire farming—as far as productivity was concerned—was not the age of enclosure before 1815, but the 1820's and 1830's.

A comparison of Table 3 (p. 59) with Table 17 shows that the increase in yields had not been regionally uniform. In 1800 there had been a close connexion between inherent soil fertility and crop yields, so that the fen soils gave the highest yields followed by the clays and the limestones. This was not so in 1851. Yields on the light lands—the limestone and gravels—had increased far more than on the clays or the fen, so that there was relatively little difference between the yields on the three types of land. The clays, in fact, now ranked slightly below the limestones, the fens slightly above (Fig. 9, p. 61). An equally important difference was between the drained and undrained clays. This is reflected in Table 17, which shows that the well-drained clays of west Kesteven gave a higher average wheat yield than the undrained boulder clays of southeast Kesteven.

To summarize then: both the literary and statistical evidence shows that there was a remarkable improvement in farm management in South Lincolnshire between 1815 and 1851, and that the pace of improvement—in terms of productivity—was far greater than between 1770 and 1815. The whole of South Lincolnshire was affected by these improvements. The critical advances seem to have been the adoption of stall feeding, the inclusion of a year of seeds in the rotation, and the liberal application of artificial fertilizers and farmyard manure. These features were common to the whole area, as was the development of tenant right. South Lincolnshire—and indeed Lincolnshire in general—was thus sharply demarcated from the grazing districts in the west and Midlands by its adoption of mixed farming with an increasing emphasis on arable land; on the other hand the practices such as stall feeding on oil-cake, tenant right and the regular growth of temporary grasses differentiated Lincolnshire from some other 'improving' areas of eastern England. Further, within South Lincolnshire regional differences remained. South-east Kesteven remained particularly backward, whilst the clays had very different problems to solve from the Heath and gravel areas. Thus, three levels of regional differentiation resulted from the

The Agricultural Revolution in South Lincolnshire

improvements of the 1820's and 1830's; first, between Lincolnshire and those parts of Britain where 'mixed' farming was not adopted; secondly, between Lincolnshire and the other regions where 'mixed' farming was adopted—such as East Anglia, the Lothians and Northumberland; and thirdly, between regions within South Lincolnshire. This latter point will be discussed further in chapter x.

CHAPTER IX

THE NEW AGRICULTURAL GEOGRAPHY

The radical changes in farm management which took place between 1815 and 1851 were not without their impact on the agricultural geography of the area. Yet as in the war period the greatest changes came in land use; farm-size structure and the pattern of landownership did not experience the re-organization which might have been expected in view of the widespread adoption of new farming methods.

It will be remembered that during the Napoleonic Wars there was a continuous increase in both the total agricultural area and the arable acreage. Most of the new arable land came from the ploughing—in many cases for the first time—of marginal land in the Interior Fens and on the lighter soils of the gravel and limestone areas. But by 1825 the maximum agricultural area had virtually been attained, so that any further increase in arable land had to come from the ploughing of permanent grassland. But this was exactly what should *not* have happened in the 1820's and 1830's if the experience of other parts of England at this time is to be any guide.[1] Grain prices—particularly wheat prices—were poor throughout most of these years, and in many parts of England farmers allowed arable land, especially that on heavy soils, to 'tumble' down to grass. Indeed this is a characteristic feature not only of the 1820's and 1830's, but of the two periods of low grain prices at the end of the nineteenth century and between the two World Wars. In both these periods the arable acreage of the country fell substantially and it was on heavy clays that the increase in grassland was most noticeable.

This might have been expected in South Lincolnshire. The clay areas of south-east and west Kesteven already had a high ratio of grassland and their arable lands were badly drained until the 1830's; and it was thus difficult for these farmers to compete with the freely drained and more easily tilled lands on the Heath. In addition the

[1] G. E. Fussell and M. Compton; G. E. Fussell, *Journal of the Land Agents' Society*.

clays gave a fair grass, and as livestock prices were generally better than grain prices there seems every reason to expect the grassland acreage in these two regions to have increased. Yet the reverse happened; not only were there few instances of land being laid down to grass, but there was a continuous increase in arable from 1820 onwards. In west Kesteven this was most marked between 1825 and 1835, and witnesses made special mention of this before the Select Committees on Agriculture. In south-east Kesteven there was a similar increase. Much of the Ancaster estate lay on the boulder clays of this region, and the estate letters between 1820 and 1850 contain repeated requests from tenant farmers for permission to plough their grassland.[1]

On the lighter soils of the Heath and the gravel soils of northwest Kesteven the situation was different. This had always been a predominantly arable area, but with much waste. Waste land had been ploughed up during the war period, when high prices justified farming even this low-yield land, but after the wars low prices prompted some farmers to allow the land to return to waste; the low yields and low prices did not cover production costs, and the soils gave a poor grassland. There is some evidence of land 'going back' on light soils in other parts of England at this time, and in the 1880's and 1930's parts of the Heath were little better than waste land. Yet it was for only a brief period immediately after the end of the Napoleonic Wars that land went out of cultivation. The remaining waste land on the Heath was ploughed up in the 1820's and 1830's, so that by the middle of the century there was little land not used for agricultural purposes in this region. The last major enclosure on the Heath was in fact made in 1823, when Mr Chaplin enclosed 2500 acres of rabbit warren on Blankney Heath and put it under the plough. In north-west Kesteven the soils were perhaps the poorest in South Lincolnshire; small pieces of 'moor' were being 'taken in' throughout the 1820's and 1830's, but gorse remained a noticeable feature of the landscape as late as the 1840's, as can be seen from the Tithe Redemption Award maps.[2]

[1] *B.P.P.* V (1833), Q. 12, 199; III (1836), 23–4; L.A.O. 2 Anc 7/4/1, 7/22/93.
[2] L.A.O. Cragg 1/1 (Leasingham, Cranwell, Hanbeck); *J.R.A.S.E.* XII (1851), 376; L.A.O. Jarvis 8, Tithe Redemption Awards.

The New Agricultural Geography

But the greatest increase in arable came in the fenland. At the beginning of the century the Townland had been mainly under grass and the interior fenlands a mosaic of intermittently cropped land with considerable stretches of poorly drained grass. The cultivation of these fens had begun before the end of the wars, particularly in the north. But it was not until the improvements to the fen drainage described in chapter VIII that the Interior Fens were finally converted to permanent arable land. This was a truly revolutionary change in land use, and it should be remembered that the same process was going on in the fenlands of Cambridgeshire, Huntingdon and Norfolk. So great was the addition to the arable acreage of England by this successful drainage that some thought the depression of wheat prices was due to the great increase in supply from the fens. Not only were the newly drained Interior Fens converted to arable but the permanent grassland of the Townland was also encroached on by the plough. These changes in land use in the fenland can be adequately traced in the contemporary literature;[1] there is also some confirmatory statistical evidence. Table 18 (p. 158) shows the quarters of wheat sold in Spalding and Boston, the two major Lincolnshire fenland markets, in selected years between 1820 and 1841. Part of this enormous increase must be attributable to increased yields. But even if it is assumed that yields doubled between 1820 and 1841, the figures still suggest a massive increase in the wheat acreage.

It seems undeniable that there was a substantial increase in the arable acreage between 1815 and 1851 and that this increase was quite different from the arable expansion of the late eighteenth century, when little permanent grassland was ploughed. Further, there was almost certainly a net increase in the arable acreage of England as a whole between 1815 and 1851 in spite of the uncertain grain prices of these three decades and the temporary local increases in grass.[2] Here then is a paradox comparable with that discussed in chapter VII. Why should the arable acreage have increased when not only were grain prices low and fluctuating but the prices for live-

---

[1] *B.P.P.* III (1836), Q. 8805, 6038–40; P. Thompson, *Collections*, pp. 371–82; *idem, History*, pp. 691–5; *J.R.A.S.E.*, XII (1851), 384.
[2] G. R. Porter, pp. 139–44; Drescher, p. 167.

stock products were generally much more buoyant? Witnesses before the Select Committee on Agriculture, which sat in the 1820's and 1830's, were nearly all agreed that agricultural distress was greater on arable than grassland because of this disparity in price movements.

TABLE 18. *Quarters of wheat sold at Boston and Spalding markets, 1820–41*

	Boston	Spalding
1820	28,112	3,479
1825	54,904	17,516
1830	75,799	23,726
1836	121,831	37,185
1841	113,409	32,763

Source: B.P.P. VIII (1836), part 1, p. 215, XL (1842), 617.

Relatively few writers have discussed this paradox. However, Levy, writing in 1911, did comment on it. He thought the explanation lay in the continuance of the Corn Laws which were specifically designed to protect the grain producer. Although the Laws failed in their purpose and grain prices continued to fall, farmers constantly expected higher prices to return and thus maintained and in many cases increased their arable acreage. 'Although neither the tariff of 1815 nor that of 1828 were able to keep the price of corn anywhere near the desired level, yet the farmers behaved as though that high price, which they supposed to be guaranteed, existed....' Thus the continued expansion of arable farming was 'based on an imaginary foundation: it resulted from an assumption which was never justified'.[1] This may be a partial explanation of the phenomenon. But it cannot be accepted as a complete explanation. Certainly in South Lincolnshire other factors were at work.

The typical production unit in the area, except on the Heath, was the small family farm. When the price of an agricultural commodity is falling, small farmers frequently react by producing more of that commodity in order to maintain their income. There seems little doubt that this helps to explain the reactions of many small

[1] Levy, p. 49.

The New Agricultural Geography

farmers in west and south-east Kesteven. Three additional factors reinforce this view. In the first place newly ploughed permanent grass gave unusually high crop yields; the farmer thus got a temporarily increased output with no extra input. In the second place farmers had yet to develop adequate techniques for grass farming. Newly sown grass took some years to come into full production. Thus if a farmer laid arable down to grass—or as was more likely let it 'tumble' down—his returns per acre for the first few years would be lower than returns either from permanent grassland or from arable. As the small farmer could not afford to wait for his inexpertly sown grasses to come into full production, he preferred to continue ploughing in spite of the falling grain prices. Lastly, in several years—particularly in 1827—sheep on the claylands of South Lincolnshire were attacked by sheep rot. In west Kesteven many small farmers lost the bulk of their flocks. As they were already in difficulties from falling grain and wool prices they were unable to restock. They had little alternative but to plough for wheat and hope that grain prices would recover.[1]

But undoubtedly the major cause of the increased arable acreage, particularly in the grassland regions of west and south-west Kesteven, and the Townland was the changing method of producing livestock. At the end of the eighteenth century the old-time grazier who had little arable was still dominant in these regions. But with the emergence of mixed farming, such graziers largely disappeared. As early as 1827 it was reported that there were hardly any old-style graziers left in south-east Kesteven. On the mixed farm stock were fed on artificial feeds and arable—root crops and 'seeds'. The decline of specialized grazing was in fact inevitable for it was a relatively extensive form of land use. Farming theory at the time held that arable was a more efficient form of land use than grazing. Thus an acre of grassland would feed so many animals per year; but on an acre of arable fodder crops could be grown which would feed nearly as many animals as the acre of grass, whilst in addition the cereals grown in rotation provided a cash crop. The increasing pressure on

[1] G. R. Allen, 'Wheat farmers and falling prices', *The Farm Economist*, VII, no. 8 (Oxford, 1954); J. Thirsk and V. Imray, *Suffolk Farming in the Nineteenth Century* (Ipswich, 1958), p. 22; *B.P.P.* IV (1833), Q. 12,199, VIII (1813–14), 243; Marrat, I; *B.P.P.* VIII (1836), part 3, Q. 7966, 8041–3; Thirsk, p. 302; L.A.O. Jarvis 8.

land in the 1820's and 1830's, when the limits of the agricultural area had been reached, necessitated this more intensive form of land use. By the time John Clarke expounded this theory in the *Journal of the Royal Agricultural Society* for 1847, based on his experience in South Lincolnshire, most of his neighbours had adopted the system. Inevitably it meant an increase in the arable acreage; even the magnificent pastures of the Townlands dwindled as mixed farming replaced the old-time grazing.[1]

TABLE 19. *The regional pattern of land use, 1838-50*

Region	Arable land as a percentage of all crops and grass	Region	Arable land as a percentage of all crops and grass
The Heath	80	South-east Kesteven	47
North-west Kesteven	70	West Kesteven (A)	42
South Holland	51	West Kesteven (B)	47

Source: L.A.O. Tithe Redemption Awards.

How did this increase in arable affect the pattern of arable and grass in the area? In Table 19 the proportion of the total agricultural area under arable in selected regions is shown. These figures are based on the data in the maps attached to the Tithe Redemption Awards. In many parishes tithe in kind had been commuted to a rent charge at the time of enclosure, but where this had not been done Tithe Commutation was carried out between 1838 and 1850. The award was based on a detailed land-use map of the parish, maps made with an accuracy never attained before and rarely matched since.[2] In South Lincolnshire there are Awards for 84 parishes. These parishes are not on the whole scattered in a random manner but occur in groups: they thus give a very accurate picture of land use in a number of regions but give no indication of the pattern in the remainder of the area. The regions referred to in the table are those delimited in chapter VI. The land-use pattern in

[1] L.A.O./T.Y.R. 4/1/103; P.R.O./H.O./42/53/99; *J.R.A.S.E.*, VIII (1847), 500; B.P.P. VIII (1828), 22-4.
[2] H. Prince, 'The tithe surveys of the mid-nineteenth century', *Agricultural History Review*, VII (1959), 14-26.

The New Agricultural Geography

the regions not covered by the Awards can be deduced from the estimates made by J. A. Clarke in 1851.

It is clear from Table 19 and Clarke's description that whilst the area under permanent grassland had been greatly reduced by the middle of the century, the *pattern* had not been substantially altered.[1] The highest proportions of grassland were still to be found in the old grazing regions of west and south-east Kesteven and the Townland. Elsewhere grass was becoming a residual feature, to be found mainly in areas unsuitable for arable. Thus the steep slopes of Lincoln Edge, between the Heath and the clay vale, remained largely under grass as they did in Arthur Young's time and still do today. In the Interior Fen grass remained only in the poorly drained parts. A noticeable example of this was Cowbitt Wash, an area to the south of the Welland which was left undrained and embanked to take the overflow waters of the Welland.

The great increase in arable land was paralleled by changes in the use of the arable. At the beginning of the century a variety of crops had been grown in the region, with a close relationship between soil type and crop selection. Thus whilst oats had been the leading crop in the area as a whole, it had by no means been the main crop in every region. But by the middle of the nineteenth century wheat had become the dominant crop in every region and 25 years later it was the main crop in every parish in South Lincolnshire, save for a few on the Heath where barley replaced it. This was a direct consequence of improvements in farm management. In the fen in the early nineteenth century oats had generally been grown in preference to wheat because of the poor quality grain produced on the peat and skirty soils. This had, however, been overcome by 'claying'. Clayed peats gave a better quality and higher quantity of wheat, and oats steadily lost its primacy in the second and third decades of the century. Not only was the newly drained land put under wheat rather than oats, but wheat replaced much of the existing oats acreage, so there was probably an absolute decline in

---

[1] L.A.O. T.L.E. 38/7, Whitfield XV/2; *B.P.P.* xxx (1834), 291*a*–298*a*; *B.P.P.* xxvii (1850), 375; *J.R.A.S.E.* vii (1846), 502; viii (1847), 116–17, 130–2; xii (1851), 380–5.

the oats acreage. This can be inferred from Table 20 which shows the amount of wheat and other grains shipped coastwise from Boston between 1815 and 1850. This by no means represents the total production of grain in the fenland, but it clearly shows the decline in 'other grains'—which were largely oats—and the increase in wheat.[1]

TABLE 20. *Coastwise shipments of grain from Boston, 1815–50*

Year	Quarters of	
	Wheat	All other grains
1815	22,275	246,343
1818	20,539	187,700
1830	34,871	114,838
1850	64,648	49,751

Source: P. Thompson, *History*, p. 351.

On the lighter soils of the Heath and north-west Kesteven there were equally marked changes in cropping. At the beginning of the century the thin soils, deficient in both nitrogen and phosphorus gave a poor wheat, and barley and oats were the main crops, whilst on the gravel soils rye was still widely grown. But the growing use of bones and farmyard manures combined with folding greatly improved the soils of these two regions and enabled wheat to be much more widely grown. By the middle of the century it was the leading crop in both regions, although on the Heath barley was a close second. In west and south-east Kesteven there was less change in crop selection, for wheat had been the main crop in both regions at the beginning of the century. However, by the middle of the century it had become the dominant crop. Maltsters increasingly found the barleys of the poorly drained clays unsatisfactory and even when these lands were underdrained in the 1820's a wet harvest still meant a poor malting barley. Permanent grassland was ploughed up to sow fodder crops, and the wheat acreage automatically increased as part of the new rotations. The decline of the sheep flocks in west Kesteven meant that the small farmer there became

---

[1] *B.P.P.* VII (1847–8), Q. 482; VIII (1836), part 1, appendix no. 4, p. 215.

The New Agricultural Geography

increasingly dependent on wheat for his income. Wheat was, as one farmer in this region put it, 'our sheet anchor'.[1]

The main change among the grain crops then was the spread of wheat, so that by mid-century it was the dominant crop in every region in South Lincolnshire. There was less uniformity amongst the fodder crops. Fodder crops—by which is meant here roots and temporary grasses—increased their relative importance in the rotation as the four course and its modifications came to be the general system of management. Perhaps the most striking change was the increase in temporary grasses. Although the 1801 returns make no mention of 'seeds' they were clearly already grown in every region in South Lincolnshire by some farmers; by the middle of the century seeds were common to nearly every farm and some farmers had already adopted a course with 2 years under rotational grasses. Grasses, unlike roots, were not greatly limited by soil type, and it seems that as much stress must be put on them as roots in accounting for the increased productivity of land in the area. Certainly they must have been a conspicuous part of the new land-use pattern.

Roots—of which turnips were still the main component—were about a quarter of the arable land on a light land farm in the middle of the century, and they were the key crop on the Heath and gravel soils. By 1851, however, the swede was beginning to replace the turnip, particularly on the light silts of South Holland. But the turnip and swede had not extended far outside the limits of 1801, for they were only successful on the lighter soils. The turnip could be grown on the thinner and better drained clays, but not with much success, and beans remained the main fodder crop. In the Interior Fens coleseed remained an important fodder crop but certainly not with the overwhelming importance it had possessed at the beginning of the century.[2]

The major grain and fodder crops still occupied the greater part of the arable land in the fenland, but by 1851 there were already signs of the specialized farming which was to be so characteristic

---

[1] *J.R.A.S.E.* IV (1843), 302; XII (1851), 339; *B.P.P.* VIII (1836), part 1, Q. 8387; L.A.O. Jarvis 8, T.L.E. 38/7.
[2] *J.R.A.S.E.* XII (1851), 339, 376–7, 378, 380, 396; Caird, pp. 183, 188; *J.R.A.S.E.* VIII (1847), 123, 126, 132.

later in the century. Woad remained locally important, but flax and hemp had by now disappeared. However, new crops had begun to be grown, some with a surprisingly modern flavour. Chicory, for example, was grown, used as a substitute for coffee, and establishments to process the crop were erected at Spalding, Holbeach and Algakirk. On the marshlands of South Holland mustard seed was quite widely grown. But the most important of the minor crops was potatoes. Although in mid-century it was still only of local importance the acreage expanded steadily from 1850 onwards, so that by 1875 a tenth of the arable land in Holland was devoted to it.[1]

In 1826 some marshland in the parish of Long Sutton was ploughed for the first time; its utilization in the following years illustrates the importance of minor crops in Holland. The figures below show the use of one acre for a number of years:[2]

Mustard seed	40%	Potatoes	29%
Wheat	26%	Beans	5%

Some of the changes in livestock farming which took place after the Napoleonic Wars have already been discussed, most noticeably the changing methods of rearing and fattening livestock and the decline of sheep in west Kesteven. But it must not be thought that all the early characteristics of livestock husbandry were swept away by the new methods. The old grazing regions retained the highest proportion of permanent grassland in the area, and it was still in these regions that sheep and cattle were fattened before being sent to market. This much is clear from Clarke's description, and it can be shown statistically when the Board of Agriculture's returns become available after 1866. In Table 21 livestock densities for the major regions are shown for 1875.

It can be seen from the table that the pattern of livestock farming which had emerged in the late eighteenth century was still present in the 1870's. The highest livestock densities were generally in the three grazing regions of the Townlands, south-east and west Kesteven; the Heath and north-west Kesteven, the regions with the highest arable proportion had the lowest livestock densities. There

[1] J.R.A.S.E. VIII (1847), 117; XII (1851), 385.   [2] Ibid. VI (1846), 502.

The New Agricultural Geography

are, however, two noticeable exceptions to this. In west Kesteven the ravages of foot rot in the 1820's had not been made good, and sheep ceased to be of much significance, although cattle were still numerous. In south-east Kesteven, which in the 1800's had been noted for both cattle and sheep, cattle seem to have declined in importance though the area retained a high sheep density. The Townland, in contrast, remained important in both cattle and sheep production.

TABLE 21. *Regional livestock densities, 1875 (per hundred acres of agricultural land)*

	Cattle	Sheep		Cattle	Sheep
South Holland	19	106	West Kesteven (A)	16	81
Skirbeck	16	86	West Kesteven (B)	18	89
Kirton	16	118	South-east Kesteven	13	110
Kesteven Fen	11	102	The Heath	8	94
Deeping Fen	13	97	South Lincolnshire	15	101
North-west Kesteven	12	90	England and Wales	18	83

Source: *Board of Agriculture Returns, Parish Abstracts, 1875.*

Comparisons between 1798 and 1875 can be little more than surmise for the livestock statistics for the earlier period do not have the reliability of those collected by the Board of Agriculture. Yet there certainly seems to have been an increase in the number of cattle in the area as a whole, for whereas in 1798 there had been 770 sheep for every 100 cattle in South Lincolnshire, by 1875 there were only 650 sheep for every 100 cattle. As the total number of both cattle and sheep in 1875 exceeded that of 1798, the changing ratio must be due to a greater increase in the total number of cattle. In addition there may have been an absolute decline in the number of sheep in certain regions. In west Kesteven this seems to have happened. It may also have occurred in south-east Kesteven; certainly the only available parish densities for the early part of the century indicate a much higher density of sheep at that time than in 1875 (see p. 80). However, this local decline in sheep numbers was probably compensated for by an increase on the arable lands of the Heath and

The Agricultural Revolution in South Lincolnshire

north-west Kesteven. The growing use of arable land for fodder and the spread of folding allowed these regions to carry more sheep than at the beginning of the century.

When the Governors of Christ's Hospital visited their Skellingthorpe estate in 1848 they noticed with surprise and regret the small number of cattle on the farms.[1] In mixed farming cattle were increasingly the centre of the system, for they not only provided the rich manure that gave high crop yields, but the shift of prices in favour of livestock products meant they provided an increasing proportion of farm income. Yet 25 years later cattle were still a relatively unimportant feature of South Lincolnshire farming. Only in South Holland did cattle densities rise above the national average (Table 21, p. 165), and by but one point. Clearly the three grazing regions of South Lincolnshire were only locally of significance as cattle producers. On the other hand only one region—west Kesteven —fell below the national average for sheep. Lincolnshire remained a great sheep county for some decades to come.

It might be thought that the transformation of farming methods and land use would have been paralleled by similar changes in farm organization; yet there seem to have been only small changes in either farm size or landownership between 1815 and 1851, or indeed in the 'golden age' between the Crimean War and the 1870's. Yet most authorities have assumed that farm size steadily increased during this period. Levy in particular supposed that the extinction of the small farm was completed in the 30 or 40 years after the end of the wars. Levy's evidence was largely literary; and it is not difficult to find descriptions of amalgamations in South Lincolnshire which would support his thesis. Thus a witness before the Richmond Commission in 1881 testified that 'when the very bad times came in 1848–52 the small farmers broke and disappeared in the way that you see leaves in the autumn'. Other witnesses before this Commission, and that of 1894–6, considered that the amalgamation of farms in Lincolnshire had been widespread before 1871, although this process had been halted by the depression. But these witnesses were often describing the circumstances of their own

[1] L.A.O./T.L.E. 38/7.

The New Agricultural Geography

locality alone and failed in addition to point out that the amalgamation of a few farms made only a small impact on farm-size structure in general. The little statistical evidence there is on changes in farm size between 1815 and the 1870's does not suggest that any radical change was going on.[1]

In Holland some writers thought that the small farmer was being slowly extinguished. John Cragg, writing in 1817, deprecated the fact that large occupiers were 'swallowing up' the small in Kirton. Farther north John Leaf wrote of Skirbeck in 1856 that

> ...formerly a great proportion of the farmers were freeholders and the number of cottagers, also freeholders, occupying a few acres of land was very considerable. But the freeholders of neither class, though still a considerable number are by no means as numerous as they once were; for as land has come into the market, a great deal of it has been absorbed by the larger farms...where even a small farm becomes vacant, a large farmer is almost certain to apply for it and to add it to an already overgrown occupancy.[2]

Certainly tenurial conditions in Holland were favourable to amalgamation. Land was rarely passed on directly from father to eldest son. Unless a farmer died intestate, the farm was divided amongst all the sons. More frequently one son took the whole farm but paid his brothers their share of the inheritance. This invariably involved selling part of the farm and so there was constantly land on the market. Yet the statistical evidence suggests that there was little change in farm-size structure. Table 22 (p. 168) compares the farm-size structure of Skirbeck in 1813 and 1870.

The table shows that there had been a slight decline in farms between 50 and 99 acres, a slight increase in those over 100 acres and no significant change at all in the proportion under 50 acres.

In Kesteven there is no equivalent to the Verdict of the Court of Sewers in Holland; however, some evidence of farm size can be obtained by comparing estate surveys in the early nineteenth century with the Tithe Awards of the 1840's and the Census returns of 1851. These records show that some amalgamation was going on. In west Kesteven, for example, it occurred in Claypole, Marston and Syston.

---

[1] Levy, p. 49; *B.P.P.* XVI (1881), Q. 56,210; XVI (1894), part 2, Q. 14,370, 14,610.
[2] L.A.O. 2 B.N.L.; P. Thompson, *History*, p. 693.

The Agricultural Revolution in South Lincolnshire

But generally this was not being achieved by the absorption of small farms by larger farms but by the amalgamation of two or three medium-sized farms. There were also examples of large farms being broken down into a number of small farms. Both these trends are illustrated in the only two parishes that can be directly compared at well-spaced intervals; Doddington, in 1811 and 1835 and Rippingale in 1804 and 1851. At Doddington two farms of over 300 acres were broken into a number of medium-sized farms, whilst at Rippingale a number of medium-sized farms were regrouped into two large farms. Similar instances of large farms being broken down occur at Cranwell and Hougham.[1]

TABLE 22. *Farm-size structure in Skirbeck Hundred, 1813 and 1870*

	Percentage of all farms over 5 acres in each size group	
	1813	1870
Over 100 acres	8	10
50–99 acres	13	11
5–49 acres	79	79

Source: Boston, H.C.C. 32/1 and 8; *Board of Agriculture Returns, Parish Abstracts, 1870.*

The 20 or 30 years of prosperity following the Crimean War might be expected to have been a period of increasing farm size, for machinery was for the first time becoming important in English farming. As has been noted earlier, Lincolnshire men considered that this period saw a great deal of farm amalgamation. Yet the only reliable statistics do not support this view at all. In 1851 the Census collected data on farm sizes and this material is available for parishes in the Public Record Office and by counties in the Parliamentary Papers. In 1871 the same data were collected but are only available for counties. Table 23 compares the farm-size structure of Lincolnshire in 1851 and 1871.

[1] *B.P.P.* xxx (1834), pp. 291a–298a; P.R.O. H.O. 107/2094–2104, 2136–8; L.A.O. Tithe Redemption Awards. See also references to Table 11 (p. 93).

The New Agricultural Geography

TABLE 23. *Farm-size structure in Lincolnshire, 1851 and 1871*

	Percentage of all farms over 5 acres in each size group	
	1851	1871
1000 acres and over	0·59	0·64
500–999 acres	3·5	4·6
300–499 acres	7·4	8·4
100–299 acres	17·9	15·8
50–99 acres	26·3	23·8
5–49 acres	44·9	47·0

Source: B.P.P. LXXXVIII (1852–3), part II, p. 597; LXXI (1873), part II, p. 127.

The table shows that between 1851 and 1871 there was a slight increase in the numbers of the smallest farms, a decrease in medium-sized farms and a slight increase in all groups over 300 acres. But there was certainly no radical change.

It must be concluded then that whilst the literary evidence suggests that there was continued amalgamation of farms and absorption of small farms in South Lincolnshire after 1815, the statistical evidence shows that both the amalgamation and the breaking up of large farms went on side by side. As a consequence, whilst there may have been a slight increase in the importance of the larger farm, there was most certainly no marked change in farm-size structure.

There seem to be a number of reasons to explain this stability. In the first place enclosure was virtually completed by 1815. At enclosure there was every opportunity for fragmented farms to be consolidated and for farmers to acquire additional land. After 1815 the layout of farms solidified. Thus the very fact that a farmer wanted to increase the size of his farm did not mean he easily could. It is true that land came on the market, but he would generally want it to be adjacent to his existing farm. This was comparatively rare.

In the second place economic conditions between 1815 and 1851 were such that there was little incentive for a farmer to increase the size of his farm. Whilst there were always applicants for vacant small farms in the bad years, it was sometimes difficult to find a tenant for a large farm. The agent of Lord Willoughby's estate wrote in 1851: 'The fact is there is a general apathy among farmers as to the taking of large farms unless these farms are in the best localities and in good condition...[they] say a man who has capital sufficient to stock such a farm...is better able to get a living by investing in something else'. In some cases agents had to split up large farms to get tenants, as a Holbeach landlord had done over 80 years before. Because wool prices were low this landlord had split a 552 acre into four 'as there are more tenants capable of using and ready to take such like than one very large farm...'.[1]

TABLE 24. *The regional pattern of farm size in 1851*

Region	Percentage of all farms over 5 acres in each size group			
	5–50 acres	50–100 acres	100–300 acres	300 acres and over
North-west Kesteven	50	17	25	7
West Kesteven (A)	47	18	27	8
West Kesteven (B)	45	13	28	14
The Heath	29	14	35	16
South-east Kesteven	33	16	35	16
Deeping Fen	34	14	31	20
Kesteven Fenland	45	24	25	6
Skirbeck	67	15	15	3
Kirton	59	17	19	5
South Holland	54	18	21	7
All Fenland	59	17	18	6

Source: P.R.O./H.O./107/2094, 2104, 2136, 2138.

Because there was so little change in the average size of farms between 1815 and 1851 it is not surprising to find that in the latter year the pattern of farm size was much the same as at the beginning of the century. Table 24 is based on the parish records of farm size collected for the Census of 1851; these statistics represent the earliest accurate and comprehensive figures on farm size for England and Wales. They underestimate the number of farms which had less than 50 acres, but this does not greatly disturb the regional distinctions.[2]

[1] L.A.O. T.Y.R. 4/1/87; 2 Anc 7/19/52.
[2] D. B. Grigg, 'Small and large farms in England and Wales; their size and distribution', *Geography*, XLVIII, no. 220 (1963), 268–79.

The New Agricultural Geography

Table 24 shows clearly the regional contrasts which had already been apparent at the beginning of the century. In the fenland the dominant production unit was the farm of less than 50 acres, and indeed a great many of the farms in this category were of less than 20 acres. The large farm was of very little importance in the fenland, only 6 per cent of all the farms being over 300 acres. However, an exception to this should be noted. As the Interior Fens were drained and farmed for the first time there were frequently opportunities to lay out new farms; and in some cases very large farms were established. This was particularly true in Deeping Fen and the fens north of Boston. There then came to be a difference in farm size between the Townland and the Interior Fens. A similar difference can also be recognized on the Tithe Redemption maps between the Townland and the Marsh, for there was a tendency for larger farms to occur in the newly reclaimed areas in South Holland. Indeed in South Lincolnshire as a whole the only very large farms—over 1000 acres—were found almost entirely on the Heath, the Interior Fens and the Marsh.

The small farm was also the most important single farm size in west Kesteven, although southwards on the clay vale large farms increased in significance; in west Kesteven (B) farms of over 300 acres were of above average importance. But it was on the Heath that the large farm really came into its own. In both the Heath and south-east Kesteven over half of all the farms were more than 100 acres in size, and in neither region were small farms more than a third of the total. The Heath was also the region *par excellence* of very large farms; units of over 500 acres were not uncommon in much of the zone stretching from Lincoln in the north to Grantham in the south-west and Stamford in the south-east. They were particularly numerous on the Heath between Sleaford and Lincoln.

It can reasonably be concluded then that there was little change in the farm-size structure or the pattern of farm size in South Lincolnshire between 1815 and 1851; and indeed a comparison of conditions in 1851 and 1871 reveals continuing stability in the structure and geography of farm size. What of landownership in this period? Sadly the records are far from complete but they suggest that there was little change either in the size of the great estates or in the status of the occupier owner. For example, if the Land Tax returns of the early nineteenth century and the list of landowners in the 'New Domesday Book' of 1873 are compared, the same great families are found to be the dominant landowners.[1] Further, those estates for which records of acreage and rent survive show little change in size. An exception appears to be the estate of the Turnor family which seems to have been considerably augmented during the nineteenth century. However, the records of landownership are at no time sufficiently comprehensive to allow any definitive conclusion to be drawn on this topic. Let it merely be said that there seems to have been little radical change in the number of major landowners.

The status of the occupier owner can be traced with a little more assurance. In Kesteven Land Tax returns survive for 1832 and can be compared with 1812 to give some measure of the effect of the depression on the small landowner. The percentage of the Land Tax paid by occupier owners in each parish in 1812 and 1832 has been compared and a number of conclusions can be drawn. In the first place, of the 183 parishes compared (there are no returns for the Soke of Grantham in 1832), 100 showed no change at all. For the most part these were parishes where there had been no occupier owners in 1812. Thus the great zone—corresponding to the Heath and south-east Kesteven—where tenants had been of overwhelming importance in 1812 continued free from occupier owners in 1832. In the remaining parishes 36 recorded an increase in the tax paid by occupier owners and 47 showed a decline. If we eliminate those parishes where the change was less than 5 per cent of the 1812 figure, then 20 parishes recorded an increase and 30 a decline. What is most significant about these changes is the absence of a regional trend. Parishes registering a decline were to be found adjacent to parishes showing an increase in both west Kesteven and in the Kesteven fenland. Thus the pattern of occupier owners in 1832 was much the same as in 1812; that is on the Heath and in south-east Kesteven occupier owners were largely absent but in many parishes in the fenland and west Kesteven they were important.

It will be recalled that between 1798 and 1812 there was an increase in occupier owners. Did the decline in occupier owners between

[1] *Return of Owners of Land, England and Wales.*

The New Agricultural Geography

1812 and 1832 merely eliminate this increase or was there a decline to a figure below that of 1798? Of the parishes which can be so compared, 20 parishes showed a higher percentage of occupier owners in 1832 than 1798, while 9 showed a higher percentage in 1798 than 1832, and 41 were the same in both years. There was no clearly defined trend, but there was a tendency for the increases made during the wars to be eliminated.

In Holland the only surviving Land Tax returns are for 1798 so that a similar analysis cannot be undertaken. However, the percentage of the Land Tax paid by occupier owners in 1798 can be compared with the percentage of the total area owned by occupier owners in the Tithe Awards. This comparison can be made for twelve parishes. Four showed no change, seven showed a decline between 1798 and the 1840's, and only one showed an increase. This suggests that the fall in the importance of the occupier owner after the wars may have been greater in Holland than in Kesteven. None the less the Tithe Awards make clear that the occupier owner was still an important element in the farming of the fenland in mid-century. Thus in the 1840's a quarter of the farmland in both Algakirk and Sutterton parishes was held by occupier owners, 44 per cent of Surfleet, nearly half of Sutton St Edmunds and just over a quarter of Fosdyke. Although there may have been a decline in the importance of the occupier owner in Holland, it still left him as a vital part of the economy.[1]

The relatively small changes in the occupier owner's status are perfectly explicable if the economic conditions of the first half of the nineteenth century are recalled. The majority of occupier owners had inherited their land before the inflationary period of the wars. They thus had neither rent to pay nor interest on loans, and were relatively immune to the depression of the 1820's and 1830's. They must have experienced a fall in their standard of living but otherwise could weather the storm. On the other hand the farmers who had bought land in the later part of the wars were much more vulnerable during the depression. In most cases they had not bought land outright but borrowed at very high interest rates. When the fall in prices came after 1815 these occupier owners were in many cases

[1] L.A.O. Tithe Redemption Awards.

unable to pay and were sold up. By the 1830's the majority of the surviving occuper owners in South Lincolnshire were those who had inherited their land. Those who had bought in the wars had gone under. Thus the pattern of landownership in 1832 was not very different from that of 1798 or earlier, before the temporary rise in occupier ownership during the affluence of the late war period. A similar sequence of events occurred later in the century. As farm income rose after 1851 many tenants bought their farms on long-term loans; when the fall in prices came after 1880 this group was broken by its interest rates, which were often higher than the rents tenants paid for farms of a similar size.[1]

[1] Board of Agriculture, *Agricultural State*, pp. 153 ff.; *B.P.P.* v (1833), Q. 12,215–19, VIII (1836), part 1, Q. 7,902; P. Thompson, *History*, p. 693.

CHAPTER X

THE CHANGING REGIONAL PATTERN

It might be thought from the preceding discussion that the most striking feature of the regional pattern of agriculture between 1815 and 1851 was its stability. In spite of revolutionary changes in methods and productivity, the major regions which were delimited for 1801 appeared to remain as strongly contrasted in 1851. For example, although the total arable acreage of South Lincolnshire increased, the pattern of permanent grassland and arable remained much the same. Similarly, the regions of greatest livestock densities were still west Kesteven, the Townlands and south-east Kesteven. The geography of farm size and landownership had been modified but not radically changed; small farmers and occupier owners continued to be characteristic of the Fenland and west Kesteven, whilst on the Heath and to a lesser extent in south-east Kesteven medium and large farms worked by tenants predominated. A comparison of rent per acre in 1815 and 1860 (Table 14, p. 99, and Fig. 19, p. 176) appears to confirm this picture of stability, for in 1860 as in 1815 the Fenland and west Kesteven had high rents per acre and north-west Kesteven, the Heath and south-east Kesteven, low rents. The only marked difference was that whereas in 1815 high rents had only been found in parts of the fenland, the improvement to drainage had led to uniformly high rents.

Yet in spite of this apparent persistence of the regional pattern, one of the most noticeable features of the first half of the nineteenth century was a decline in regional differentiation. It will be clear from chapter VIII that there was a far greater uniformity in farming methods in 1851 than there had been at the beginning of the century, and this had its impact on the agricultural geography. Two broad factors were responsible for this diminishing differentiation. First, better farming practices made farmers less dependent on the inherent characteristics of their soils. Secondly, the rate of agricultural change between 1815 and 1851 varied from region to region.

The Agricultural Revolution in South Lincolnshire

The net result was to even out farming differences between the major regions. This can be clearly demonstrated by discussing the changes which took place in rent per acre after 1815.

In Fig. 20 the assessments to the County Rate have been used to show changes in rent. This local rate was raised on an assessment

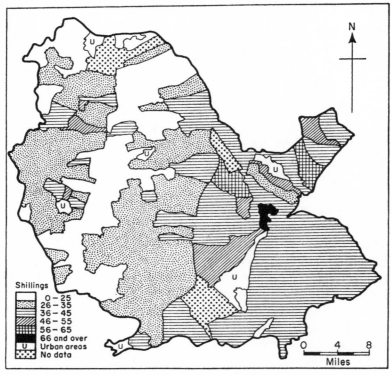

Fig. 19. Rent per acre in 1860.

based on the annual rent of land and buildings in each parish.[1] Parochial assessments survive for every parish in 1815. The changing economic conditions of the 1830's prompted a revaluation of Holland in 1840, but the Kesteven parishes were not revalued until 1847. The inclusion of the rent of 'buildings' in the parish totals makes

[1] Grigg, *Transactions and Papers*, 1962, pp. 92–3.

The Changing Regional Pattern

these figures a less accurate indicator of the rent of agricultural land than the Income Tax assessments; but it seems reasonable to assume that the differences in the assessment between 1815 and 1840–7 were due more to changes in the value of land than of buildings.

The map (Fig. 20) shows that there were marked regional differ-

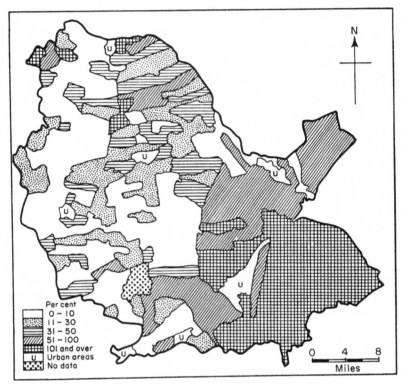

Fig. 20. Percentage change in rent per acre, 1815 to 1840–7.

ences in the rate of rental increase. The highest increases came in the fenland and on the Heath, the lowest in west and south-east Kesteven. There were a number of reasons for this.

First, the improvements in agriculture at this time were still mainly in arable farming and associated with the spread of the Norfolk system. But the Norfolk four course, whilst easily adopted without any modification on the light lands of the Heath and north-west Kesteven proved impracticable on the clay soils. Roots were rarely successful, underdrainage was difficult and folding thus unwise. Between 1815 and the 1840's most of the Heath was put under the four course and heavy expenditures made in improvements, so that rents rose considerably. On the clays, however, farmers found no completely satisfactory system of farming to replace the old ways, and rents consequently rose much less than on the Heath. The Norfolk system could not be adopted in the fenland but this was no barrier to progress there, for the great fertility of fenland soils made a strict rotation less necessary. The increases in rent there were primarily due to the successful improvements in surface drainage which were carried out in the 1820's and 1830's, so that the greatest increases in rent came in the Interior Fens.

Secondly, whilst grain prices had remained high—that is before 1814—the clays with their inherent natural fertility could profitably produce grain in competition with the light lands, even though the rate of farming improvement was already greater in the latter regions. But when grain prices fell after 1814 the clays found it increasingly difficult to compete. There was 'a fierce but silent contest carrying on between the productive lands of England and the unproductive', declared Joseph Saunders in 1836; it was a struggle that the light lands won. Clay farmers found their heavy-textured soils expensive to cultivate and costly to underdrain. A late spring had disastrous effects on yield and no means had been found of eliminating the bare fallow. Light land farmers, on the other hand, had much lower production costs, for their soils did not need underdrainage and were more easily tilled. They could concentrate expenditure on fertilizers with a resultant increase in yields. As a result production costs per bushel moved steadily in favour of light land farmers who thus found themselves much less vulnerable to low grain prices than the clay farmers.

Thirdly, these advantages were reinforced by changes in farm price structure in the 1820's and 1830's. Whilst all agricultural prices were low, malting barley and wool generally held up better than the price of other products. These were products of the Norfolk system and particularly of the limestone lands, and thus formed a higher proportion of farm income there than on the clays.

The Changing Regional Pattern

Clay farmers were doubly unfortunate for the clays gave a malting barley increasingly unacceptable to maltsters and the foot rot of the 1820's destroyed many of their sheep. As shown earlier (pp. 162–3) they became increasingly dependent on wheat prices.

Thus both costs and prices shifted in favour of the light land farmers in the 1820's and 1830's. But the fenland's competitive position improved in relation to either the heavy or light land regions. Fen soils were friable and easily tilled, and whilst the surface drainage was secure they did not generally need underdrainage. Further, their fertility was naturally far greater than most other soils so that little application of fertilizer was necessary. The high fertility and good texture allowed a wider range of crops to be grown than elsewhere in South Lincolnshire. Thus after the surface drainage had been improved fen farmers had lower production costs per bushel than in Kesteven and also a greater flexibility in their choice of crops. Not surprisingly rents rose considerably between 1815 and 1840.

As a result of these factors rental increases were least in the two clay regions—south-east and west Kesteven—and greatest in north-west Kesteven, the Heath and the fenland, particularly in the Interior Fens. As the highest rents per acre in 1815 had been in the two clay regions and on the Townland, the differential rate of rental increase after 1815 evened out regional rents per acre, as can be seen from Table 15 (p. 102). Although in 1860 west and south-east Kesteven still remained more highly rented than the Heath or north-west Kesteven, the difference was much less than in 1815. The persistence of a relatively high grass ratio and a predominance of small farms explains the continued high rents in west Kesteven; but in terms of the economic and technological conditions of the time, west Kesteven was almost certainly over-rented. Rent per acre in the fenland remained in 1860 still well above the rest of South Lincolnshire, but there was now less difference between the Interior Fens and the Townland on the one hand, and between the fenland and the rest of South Lincolnshire on the other. In 1815 rents in the drained parts of the fen had been four times the rent of the Heath; in 1842 average rents in the fenland were still double those on the Heath but by 1860 the difference had fallen to 80 per cent.

The Agricultural Revolution in South Lincolnshire

The differential rental increases were the main reason for the increasing uniformity of rent per acre in 1860, but there were other more general factors which contributed to this diminishing differentiation. At the beginning of the century farming had been regionally specialized, and each region drew its income from a limited number of products. But the spread of mixed farming diversified regional production and meant that each major region was dependent on the same range of products. In so far as farm prices affected rent per acre this led to increasing uniformity. A corollary of this was the diminishing contrast in land use (see p. 181). As there was a difference in the rent per acre of permanent grass and arable, this factor also tended to reduce rental differences. The improvement of communications within the area led to the same result. At the beginning of the century there were certainly regions—particularly south-east Kesteven and South Holland—where isolation reduced rents. The spread of better roads, waterways and later railways improved the accessibility of these regions and helped to even out the rental pattern. Lastly, the improved farming methods led to a decline in dependence on the 'inherent characteristics of the soil' and this in turn reduced regional differentiation.

The changes in rent per acre in fact summarize many of the changes in production methods, land use and regional productivity, and it remains in this chapter only to show how the regional differentiation of these elements of the agricultural geography also declined between 1815 and 1851.

In 1800 there had been very marked regional contrasts in land use. An unknown—but quite high—proportion of the area was in what would now be called rough grazing, particularly in north-west Kesteven, the Heath and the Interior Fens. Much of this had been ploughed by 1815, and certainly by the middle of the century waste survived in but a few limited localities. More important was the changing proportion of arable and grass. Whilst the old grazing system persisted, west Kesteven, south-east Kesteven and the Townland had between 70 and 80 per cent of their agricultural land under grass. Yet elsewhere grass was not a conspicuous element of the landscape. In north-west Kesteven, the Interior Fens and the Heath

The Changing Regional Pattern

permanent grass could not have occupied more than a third of the agricultural area. These differences were exaggerated between 1770 and 1815 because the greatly increased arable acreage came from the ploughing of rough grazing in these already predominantly arable areas. Conversely, after 1815 the new arable came mainly from the ploughing of permanent grassland in the three grazing regions as different methods of producing livestock were adopted. As a result there was a diminishing regional differentiation in land use (Table 25).

TABLE 25. *The regional pattern of land use, 1800–75*

	Percentage of agricultural land under arable		
	About 1800	1838–50	1875
The Heath	c. 70	80	88
North-west Kesteven	c. 60	70	83
Kesteven fenland	No estimate	62	74
South Holland	25–35	47	69
West Kesteven	25–35	45	58
South-east Kesteven	20–30	47	56
Range each year	35–50	35	32

Source: L.A.O. Tithe Redemption Awards. *Board of Agriculture, Parish Abstracts, 1875*. See also note on p. 67.

In Table 25 the estimates for 1800 cannot be given too much credence; yet there seems little doubt that there was a most striking difference at that time between, for example, the grass acreage in south-east Kesteven, where contemporaries thought at least four-fifths of the agricultural land was under grass, and the Heath, where the land not under arable was mainly waste. By 1850 the arable acreage in every region had increased, but the regional contrasts had diminished. By the time the Board of Agriculture's figures become available, the arable acreage had continued to increase, but the regional differentiation had further declined. When it is recalled that a higher proportion of the mainly arable regions—such as the Heath—were under temporary grasses than the regions where permanent grass survived in some quantity—such as southeast and west Kesteven—the declining differentiation in land use appears even more clearly.

The Agricultural Revolution in South Lincolnshire

At the beginning of the century no more than a minority of farmers followed a strict rotation and few devoted much of their land to feed crops. By mid-century a four- or five-course rotation was followed in most regions and this had its effect on the use of the arable land. Table 26 compares the proportion of the total arable land in the major regions allocated to grains and fodder crops in 1801 and 1875. It should be noticed that the 1801 crop returns made no mention of artificial grasses; nor are they included in 'fodder crops' in part B of the table although they are included in the total arable. Thus, the proportionate share of the total arable land in 1801 under fodder crops is exaggerated in comparison with 1875. The table shows that a greater proportion of arable land was under grain crops in 1801 than 1875 in every region. More important here is the fact that the difference between regions in the relative share that grains and fodder crops occupied was greater in 1801 than 1875.

TABLE 26. *Percentage of the total arable acreage in each region under grains and fodder crops, 1801 and 1875*

	A 1801		B 1875	
	Grains	Fodder	Grains	Fodder
South Holland	87	14	54	26
Kirton	77	21	56	26
Kesteven fenland	76	21	57	27
North-west Kesteven	66	24	52	24
The Heath	71	27	52	24
West Kesteven	69	27	56	24
South-east Kesteven	73	24	53	26

Source: P.R.O./H.O./67 Lincs.; *Board of Agriculture Returns, Parish Abstracts, 1875*.

In 1801 the proportion of the arable acreage under grain crops varied from 87 per cent of the total arable to 66 per cent, whilst in 1875 the variation was only between 52 and 57 per cent. Similarly in 1801 fodder crops varied between 14 and 27 per cent, in 1875 only between 24 and 27 per cent.

By 1851 there was also less diversity in the type of grain crop

The Changing Regional Pattern

grown. Reference to Fig. 10 (p. 73) shows that whilst there was certainly a tendency to regional predominance of one crop—such as oats in the fens and wheat on the clays, there was none the less a great diversity in crop selection. Thus, in 1801, 25 per cent of all the parishes for which records survive had wheat as the leading crop, 38 per cent barley and 37 per cent oats. By 1851 John Clarke recorded that wheat was the leading crop in every region in South Lincolnshire, and by 1875 wheat was the main crop in 90 per cent of the parishes for which records were collected. The consistently higher price which wheat fetched compared with oats or barley accounted for the spread of the crop, but its successful cultivation was only made possible by the general adoption of artificial fertilizers and organic manures. Barley was the only competitive crop and it was the leading crop in a number of parishes on the Heath where the limestone soils gave an excellent malting barley.

With the fodder crops, however, diversity still prevailed. Mangolds and swedes were grown as well as turnips, and in the light land regions these three root crops between them occupied a fifth of the arable land. Roots had also spread on to the clays, although they were rarely of much significance. Cabbages were also being grown as a fodder crop. In the fenland rape was still grown but on the lighter silts roots had grown in significance whilst beans were found on the heavier silts. The only tendency to uniformity among the green crops was the spread of temporary grasses. On the light lands they occupied between one-fifth and one-quarter of the arable land, and on the clays they were up to 15 per cent of the land. Only in the fenland did they remain unimportant. The spread of temporary grasses was due to two factors: first they were an accepted part of mixed farming and the Norfolk course; equally important their growth was less restricted by soil type.

It is less easy to compare the pattern of livestock farming in the middle of the century with the 1800's because of the lack of reliable statistics in the earlier period, but again there does seem to have been a decline in regional differentiation. At the beginning of the century cattle and sheep were to be found in every region, but livestock densities were particularly high in the three specialized grazing regions. In 1875 these three regions still had the highest overall

The Agricultural Revolution in South Lincolnshire

TABLE 27. *Regional livestock densities, 1798 and 1875*

	1798			1875	
	Cattle	Sheep		Cattle	Sheep
Bourne and Stamford	11	22	South Holland	38	6
Lincoln	17	26	Kirton	12	18
Grantham	39	7	North-west Kesteven	−12	−10
Sleaford	0	34	West Kesteven	21	10
Spalding	−71	−61	South-west Kesteven	−7	−15
Boston	11	36	The Heath	−75	−6
Average deviation	24	31	Average deviation	27	11

Sources: L.A.O. K.C.C. Deposit; *Map and Schedule of Live and Dead stock, 1798*; *Board of Agriculture Returns, Parish Abstracts, 1875*.

livestock densities, especially for beef cattle, for pastures were necessary for fattening. However, there was probably less difference between the sheep densities of these three regions and the remainder of the area. First, the sheep rot of the 1820's had permanently reduced the numbers in west Kesteven. Secondly, the ploughing of the Interior Fens had reduced the summer grazing available to Townland farmers and possibly led to a reduction of sheep numbers. Thirdly, the improvements in light land farming had permitted a higher sheep density to be carried there. Some statistical evidence can be put forward to support these suppositions. Table 27 shows the percentage deviation of the cattle and sheep densities of each major region in 1875 from the average density; and the livestock densities in the Military Subdivisions of South Lincolnshire in 1798 are similarly expressed as deviations from the mean density for that year. These subdivisions, with the exception of Spalding and Boston, which correspond to southern and northern Holland respectively, bear no relation to the agricultural regions of the area. Table 27 shows that regional variations in sheep densities were greater in 1798 than in 1875, and this confirms the surmise made above. The regional concentration of cattle in 1875 was if anything higher than in 1798, although it should be remembered that the regional boundaries for the two years are different. The reason for the declining differentiation of sheep numbers and the continued differentiation in cattle production is simply that whereas a sheep flock could be kept on temporary grasses, a small amount of permanent grassland and roots, cattle needed good pastures for fattening.

Lastly, the changing pattern of regional productivity must be discussed. It should be remembered that in the 1850's as in the 1800's there could be striking differences in crop yield between two adjacent farms where the soil type and the farming system were much the same. Yet there was undoubtedly a marked *regional* difference in crop yields at the beginning of the century, largely due to the generally backward farming techniques and the consequent dependence of crop yields on the inherent characteristics of the soil (Fig. 9, p. 61, and pp. 60–1). But in the 1820's techniques were worked out whereby the deficiencies of soils could be modified by careful rotations and the use of fertilizers. Thus, for example, when Arthur Young visited the Heath wheat was the main crop only where the open fields persisted. The low nitrogen content of the soil and a phosphorus deficiency meant that these soils gave a very low yield of wheat and it was more profitable to grow a less demanding crop like oats, or barley which gave indifferent yields, but was of good quality for malting. The introduction of folding and the use of bone dust enabled wheat to be grown satisfactorily by the time Philip Pusey visited the same area in 1842.[1] The light soils—the limestone and gravel soils—were more responsive to fertilizers than the clays, which already had relatively high yields in 1815. On the clays fertilizers were of little use unless the land had been underdrained, and in the 1830's there were marked differences between wheat yields on drained and undrained clays. In the fenland there were no prodigious increases in yield such as were found on the Heath. Yields in the 1850's were probably below those obtained when the newly drained peat soils were first ploughed. But the practice of 'claying' had greatly improved the quality of fen grains which hitherto had a low reputation as a bread grain. The net result was that by 1851 the average yield on the Heath had risen above those of the clays, in spite of these soils' naturally high nitrogen content (Table 17, p. 152). But there were far smaller differences in average yield between different soil types than there had been in the late

[1] *J.R.A.S.E.* IV (1843), 302.

eighteenth century (Tables 2, 3 and 17, pp. 36, 59, 152). After 1851 the general level of farming improved only slightly after the rapid changes of the 1820's and 1830's. Estimates of the average yield of wheat for various districts in South Lincolnshire were presented as evidence before the Richmond Commission in 1881, and they show that not only was the average yield little above that in 1851 but that there was still little difference between one region and another.

TABLE 28. *Wheat yields in South Lincolnshire, 1870–8*

District	Yield in bushels c. 1870–8
Grantham district I	28
Grantham district II	30
Lincoln district	26
Mid and South Lincoln	30
Grantham district III	30
Sleaford district	30
Stamford district	36
Boston	40
Holbeach	32

Source: B.P.P. XVI (1881), 375.

Table 28 shows that the yield of wheat was slightly higher in the fenland than the rest of South Lincolnshire, but otherwise there was little regional variation in crop yields.

It is possible to conclude then that, whilst the strongly differentiated soil types meant that a marked regional pattern persisted through from 1800 to the 1870's, the progress of farming technology reduced a farmer's dependence on environmental conditions and thus led to a declining regional differentiation. This supports the speculations made by Theodore Brinkmann about regional change. He wrote: 'As improvements in technology level out differences of agricultural intensity, they increase the importance of natural differences. But the organic-technical improvements, on the other hand, tend to lessen the effects of natural differences.'[1] By technology Brinkmann meant improvements in transport, by 'organic-technical' improvements he meant farming methods which increased yield per acre.

[1] E. T. Benedict, H. H. Stippler and M. R. Benedict, *Theodore Brinkmann's Economics of the Farm Business* (Berkeley, California, 1935), p. 49.

The Changing Regional Pattern

It is important to notice that the improvements to roads and waterways before 1850 did not give South Lincolnshire any great locational advantage over other areas; nor did internal improvements benefit any one region within the area. Thus until the middle of the nineteenth century location had relatively little effect on South Lincolnshire farming. On the other hand the spread of a uniform farming system and in particular the advances made in rotational practice and the use of fertilizer reduced the importance of 'natural differences'—in this case soil type—and made for greater uniformity in the agricultural geography of the area.

CHAPTER XI

SOME CONCLUSIONS

Perhaps the most striking feature of agricultural change in South Lincolnshire between 1770 and 1851 was the interrelationship between the rate of increase in productivity and the decline in regional differentiation. From 1770 to 1815 the contrasts between the major regions increased. The elimination of the open fields meant that soils could be utilized for the crops most suited to them. This is clearly demonstrated by considering the cropping of the surviving open-field parishes in 1801. In that year barley or oats were the most common crops on the thin soils of the limestone Heath, except on the open fields where custom required wheat to be grown. Similarly, whilst turnips were the leading crop in the enclosed light soil areas, beans were the main fodder crop in the open fields near Market Deeping, although these sandy loams would have given excellent root crops. The progress of enclosure also emphasized the differences between the arable and grazing districts. The grazing districts were mainly in areas of heavy soil which had been largely enclosed by the middle of the eighteenth century. In 1770 most of the surviving commons and waste were in areas where arable was already the major form of land use. The high prices of the war period prompted the enclosure and ploughing of this land, whilst the grazing districts remained largely untouched by the plough.

In 1815 there were also regional contrasts in crop yields, largely because of inherent differences in soil fertility. Although there was a slow diffusion of new techniques between 1770 and 1815 there was hardly any advance in general productivity, and regional contrasts persisted. The reasons for this were various. Investment by landlords went mainly into enclosure and fen drainage, which expanded the agricultural area but did not increase yields. Whilst marginal land remained to be ploughed there was little incentive for farmers to adopt methods which were still unfamiliar and therefore risky. The very fact that farmers could make profits easily whilst grain

Some Conclusions

prices were so high militated against innovation; nor did tenants with neither long leases nor tenant right have the security of tenure which was necessary before they would make any substantial investments. Lastly, Lincolnshire landlords, unlike their Norfolk counterparts, showed little initiative in sponsoring the new methods.

Thus in 1815 South Lincolnshire remained an area of marked regional contrasts not only in land use and productivity but in farm size, landownership and rent per acre. At the same time the level of productivity was low and had made little advance since 1770.

After 1815 there were rapid and widespread changes. The adoption of mixed farming led to a reduction of the pastures which had been the 'glory of Lincolnshire' and there was an increasing uniformity of crop selection. At the same time productivity rose at a hitherto unparalleled pace, average wheat yields doubling in 30 years. There were a number of reasons for this. Demand for agricultural products continued to rise as the English population increased; the elimination of waste prevented any further increases in output from simply expanding the agricultural area. Prices were low enough to persuade farmers that increased efficiency would lower costs per bushel, but not so low as to discourage investment altogether. Landlords, in contrast with the period before 1815, were prepared to encourage improvements and tenants were given security by the development of tenant right. Now the improvements of this period were most rapidly adopted on the Heath and in north-west Kesteven, the regions which had been least productive in 1815; progress on the clays was less striking. As a result by 1851 regional differences in yields had diminished and this declining differentiation was reflected in other features of the agricultural economy.

How does this pattern of change compare with the experience in other parts of England? To early writers the period of most rapid technical change was between 1760 and 1815; it was argued that the new methods were impracticable whilst the open fields survived, and the period of Parliamentary enclosure was assumed to have been one of rapid adoption of new methods. This view has since been questioned. First, it has been pointed out that the mere fact of enclosure did not necessitate the adoption of new methods, it merely permitted them.¹ In the second place there is a growing body of information which suggests that many of the practices of the New Husbandry were well established in parts of England before the era of Parliamentary enclosure. The critical period of change then, may have come before the mid-eighteenth century. Thus in Wiltshire

> ...the agricultural revolution, in the farming countries now under notice, was the achievement of the 16th and 17th centuries, and more particularly of the period from 1575 to 1675. There was considerable improvement both before and after this period, but it was during these hundred years that all the basic problems arising from the vicious circle of medieval agriculture by which all improvement was impeded by a shortage of feed and fodder were finally solved in both theory and practice.²

In parts of East Anglia the use of turnips and other new methods have been shown to have been in use at least as early as the seventeenth century. Neither Turnip Townshend nor Coke of Holkham can any longer be regarded as great innovators.³ Many historians then doubt whether 'revolution' is a suitable word to describe the events of the eighteenth century; indeed some have written of the 'so-called' agricultural revolution.⁴

This interpretation does not at first sight seem to correspond to the sequence of events in South Lincolnshire, where it seems indisputable that the great leap forward in productivity came after 1815. But this may well be the result of a differing interpretation rather than any real difference. One of the most characteristic features of agriculture at any time is the lag between the introduction of a technique and its general adoption. At almost any period of English agriculture a contrast can be found between the best practices of a region and the general level of farming. Clearly some farmers in every region in South Lincolnshire had adopted the methods of the New Husbandry by the 1790's; and indeed may well have done so at a much earlier date. However, if the phrase 'agricultural revolution' is to have any meaning it must surely apply not simply to an increase in agricultural output by expanding the agricultural area or to the experiments of a few farmers but to a

[1] T. H. Marshall, 'Jethro Tull and the New Husbandry of the Eighteenth Century', *Economic History Review*, II, no. i (1929), 60.
[2] *The Victoria County History of Wiltshire*, IV (London, 1959), 57.
[3] Kerridge; Parker; Plumb. [4] Thirsk, p. 17; Youngson, p. 121.

Some Conclusions

general and rapid increase in the productivity of the area under consideration.[1]

On these grounds then, it can reasonably be contended that the 'agricultural revolution' in South Lincolnshire occurred in the 30 years after the end of the Napoleonic Wars, when the general adoption of the new methods doubled average grain yields. This, of course, is not to minimize the events which preceded this period. The prerequisite conditions of this upward surge were being established over a very long period, and a number of phases of change can be detected. Thus between about 1760 and 1815 the open fields were eliminated and the maximum agricultural area virtually attained by the reclamation of heath and the drainage of fenland. Investment was undertaken primarily by landlords, for neither long leases nor tenant right existed to give tenants the security they needed. There may have been a quickened diffusion of new methods, but the majority of farmers carried on in traditional ways and general productivity did not rise. This phase stands in sharp contrast to the years following the end of the wars, when mixed farming changed the system of farming, tenant right was introduced and productivity rose sharply.

What of the years before 1760 and after 1850? Clearly, during a long period before 1760 changes were going on in the agriculture of South Lincolnshire. The new methods were entering the area for the first time. It may be too that the late seventeenth century and the early eighteenth century were the critical years for changes in the pattern of landownership and farm size. Certainly the evidence presented in chapter v suggests that, in common with much of the rest of England, the occupier owner was a relatively unimportant feature of the economy by the end of the eighteenth century. It may be that he was more important in the seventeenth century and had since been eliminated. Similarly with farm size and the existence of large estates; there does not seem to have been any critical change in the average size of the unit of control or the unit of production in South Lincolnshire between 1770 and 1850. From this it can be suggested that possibly neither was a critical factor in

[1] E. D. Ross and R. L. Toritz, 'The term "Agricultural Revolution" as used by Economic Historians', *Agricultural History*, XXII, no. 1 (1948), 32–8.

the accelerated adoption of new techniques. On the other hand it could be argued that the establishment of larger farms and larger estates had in fact been carried out before 1770; recent work in other parts of England suggests that this may be so.[1]

After 1850 a further phase can be distinguished. Between 1851 and the 1870's there was no significant increase in the average yield of wheat in South Lincolnshire. However, it may be that this period saw the beginnings of a rise in output per man. It should be remembered that the New Husbandry required a greater labour force than the old traditional methods of farming. In South Lincolnshire, for example, 'gangs' of women and children were employed to remove the stones from fields on the Heath before cultivation. This additional labour was supplied by the increase in population which took place probably from the 1780's and by the workings of the Poor Law, by the development of the gang system and by migrant labour in the harvest season, mainly from Ireland. Labour-saving machinery was of little importance before 1850. The drill accomplished tasks more efficiently rather than saving labour, as did the improved ploughs. Admittedly the introduction of threshing machines and steam engines in the 1830's and 1840's may locally have caused unemployment, but it was not until after 1851 that mechanization began to have any substantial effects on farming. It was, of course, after 1851 that the English rural population first showed signs of decline. At any rate, it seems likely that 1770, 1815 and 1850 are significant breaks in the graphs of output in South Lincolnshire. Total agricultural output probably rose steadily until 1770, and then sharply until 1850. Output per acre hardly rose at all between 1770 and 1820, rose steeply between 1820 and 1850 and then flattened out again. Output per man must have been relatively stagnant from 1770 until the 1850's and 1860's, and then began to rise. This, of course, is speculative, but the few facts on productivity which exist support the assertion.

It would, of course, be dangerous to assume that this pattern of change was common to the whole of England. In the first place it should be remembered that the 'agricultural revolution' affected primarily the 'corn counties' of eastern England; in the Midlands

[1] Mingay, *Economic History Review*, XIV.

Some Conclusions

and the west grazing remained the main activity. Even within the zone of arable farming which stretched from Northumberland southwards through Lincolnshire and then south-west to Dorset local conditions varied greatly and thus the sequence of change was probably different. For example, only a few counties in this area had either long leases or tenant right; nor did tenant right in those counties where it existed necessarily have such a beneficial impact on farming methods as it did in Lincolnshire.[1] Similarly soil conditions varied from county to county within the arable area, although most counties of eastern and southern England were characterized at least partly by an alternation of heavy and light soils. Compared with the Midlands or the west of England, farms were predominantly large in eastern England, except in the fens and around London, but as in South Lincolnshire there was a striking regional variation within each county. The importance of the large landowner also varied from county to county and this may well have influenced the rate of technical change.[2] On the whole then it would be rash to assume that conditions were sufficiently uniform throughout arable England to presuppose a rate of agricultural change similar to that in South Lincolnshire. None the less at least two previous writers have suggested that the period after 1815 may well have been critical in English agricultural history. The most recent, A. H. John, has written: 'It was not until the fall of prices after 1815 forced the intensive use of imported manures and cattle food that the second great step forward was taken by the specialist grain areas of England.'[3] Clearly the supposition that the post-Napoleonic period was crucial could not be substantiated without a close examination of the general level of farming practice in eastern England at different periods between 1750 and 1850. Perhaps the most profitable way to investigate this problem would not be simply to survey the several counties of eastern England, but to examine the progress of different farming regions. Thus a comparison of Lincoln Heath and the Yorkshire Wolds suggests a very similar sequence of change; it may

[1] Caird, pp. 504–9; *B.P.P.* VII (1847–8), Q. 1129, 7844.
[2] Grigg, *Geography*.
[3] F. C. Dietz, *An Economic History of England* (New York, 1942), pp. 402–3; A. H. John, 'The Course of Agricultural Change, 1660–1760', in L. S. Pressnel (ed.), *Studies in the Industrial Revolution* (London, 1960), p. 154.

well be that a comparison of the clay areas of eastern England would also reveal a common development.

The contention that the years after 1815 were critical in the course of farming productivity in eastern England as in South Lincolnshire must remain non-proven. What of the second part of the equation, that regional differentiation declined? To show that this was so would clearly require a study of the agricultural geography of a large area, and here one can only speculate. Certainly High Farming affected primarily eastern and southern England, for it was here that physical conditions were most suited to grain growing. In the Midlands and the west pastoral activities remained dominant, and these regions were less affected by the spread of 'mixed' farming. Equally certainly there were close parallels by mid-century between the farming of the light land areas of eastern England, and to a lesser extent between the clay regions within the arable counties. But this does not prove that between the beginning and the middle of the nineteenth century there was a declining differentiation between light lands on the one hand and heavy lands on the other, as there was within South Lincolnshire. Sadly the only statistical means of measuring this, by comparing rent per acre at different periods, is not easily applicable to eastern England as a whole. Rent per acre can be calculated from the Property and Income Tax assessments of 1815 and 1860 with a reasonable degree of accuracy. But the single figure for a county will conceal the differences between clay and limestone, sands and fen, and so forth, which are sought here, for county boundaries pay little regard to soil regions. Nevertheless, the available data are shown in Figs. 21 and 22. In 1815 the highest county rents per acre were to be found in the Midlands, Lancashire, Cheshire and Somerset, and to a lesser extent near London. In these counties high rents were a function of a number of factors, particularly the prevalence of small farms, nearness to large markets, and a predominance of either grazing land or intensive arable farming.[1] The counties with the lowest rents were to be found in areas which were both remote and possessing a high proportion of poor soil: Wales, the north and the south-west.

[1] D. B. Grigg, 'An index of regional change in English farming', *Transactions and Papers of the Institute of British Geographers* (London, 1965), no. 36.

Some Conclusions

Most relevant is the condition of eastern England: in the area where the Norfolk system was most far advanced in 1815, rents per acre were little above average.

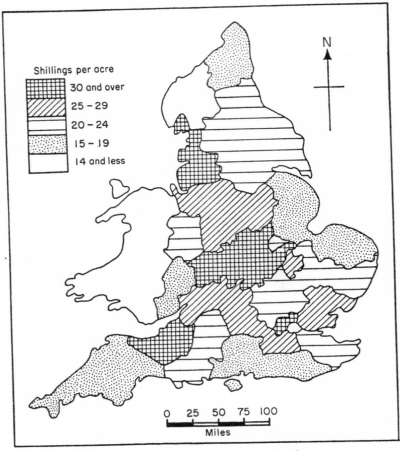

Fig. 21. England and Wales: rent per acre in 1815.

Between 1815 and 1860 the average rent per acre of land in England rose 21 per cent. Fig. 22 shows that the highest increases, 10 per cent and more above the national average, were confined to eastern England and Wales. In Wales this was due largely to two factors. First the continued enclosure and reclamation of land, and secondly the connexion of this hitherto remote and isolated area with industrial England after the coming of the railways. In eastern England, on the other hand, increases were a result of the spread of

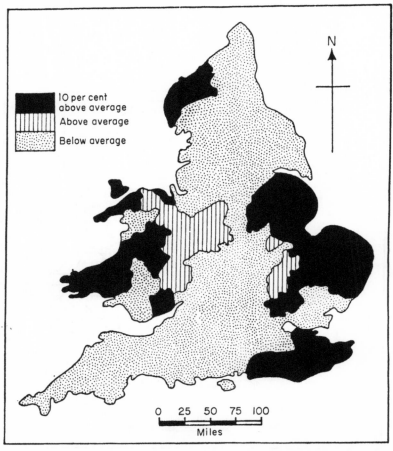

Fig. 22. England and Wales: percentage increase in rent per acre, 1815–60.

mixed farming and the heavy investment in High Farming made in this period in this region. The result was to diminish the regional differences in rent per acre in England and Wales as a whole. This is clearly shown in Table 29. The average rent per acre of the five highest rented counties and the five lowest rented counties is shown

Some Conclusions

for four selected years. The ratio between them steadily diminished from 1815 to the eve of the First World War. It could be argued that this was due to the wider diffusion of technical knowledge over the century, and hence a declining dependence on inherent soil fertility. Unfortunately neither the maps nor the table indicate whether regional differentiation *within* eastern England was declining; the experience in South Lincolnshire cannot be assumed to have been necessarily characteristic of this period.

TABLE 29. *Rent per acre in England and Wales, 1815–1912*
(shillings per acre)

	A Average rent per acre of five highest rented counties	B Average rent per acre of five lowest rented counties	Ratio A/B
1815	33·6	6·4	5·2
1860	36·6	8·0	4·5
1878	43·0	10·0	4·3
1912	32·4	9·8	3·3

So far in these conclusions the concern has been with agricultural *change*. Yet undoubtedly a striking feature of the period 1770–1850 in South Lincolnshire was the lack of change in either the structure of land ownership or farm size. It must be admitted that the statistical evidence on farm size and the size of estates in South Lincolnshire is far from conclusive; none the less it does suggest that there was little radical change in these years. This, of course, is not at all what early writers considered was the natural sequence during the agricultural revolution. Many late eighteenth-century writers—Young among them—thought that technical progress was impossible—or at least slow—whilst small farmers persisted, and that the spread of the large farm was a necessary prerequisite for technical progress. He may well have been right if a small farmer is thought to be a cottager with a scrap of land; but this type of 'farmer' even before enclosure must have been insignificant in the overall picture of farm-size structure. A small farmer can be reasonably regarded as one occupying more than 5 acres and less than 50.

The Agricultural Revolution in South Lincolnshire

There seems to have been no radical diminution of the numbers of this type of farmer in South Lincolnshire in the first half of the nineteenth century; yet at the same time productivity rose strikingly.

This is not difficult to explain. The economic advantages of large farms relate to such matters as bulk purchase and mechanization. Between 1770 and 1850 the type of technical advances which were raising yields—the use of fertilizers, oil-cake, new rotations and so forth—were as easily adopted on a small farm as a large farm. Labour saving was not a prime aim of the farmer—except incidentally—and mechanization was of little importance. It was only after 1850 that the economies of scale, which are undeniably possible on a large farm when machinery is introduced, became operative. It seems then that within South Lincolnshire the 'agricultural revolution' was attained without any substantial change in farm size, unless this in fact had occurred before 1770.

This conclusion has some relevance not only to the rest of England at that time, but to other parts of the world at the present. One recent writer has suggested that the amalgamation of farms in late eighteenth-century England has been exaggerated, and that it may be in the early eighteenth century that an increase in farm size occurred.[1] Pertinent here is the dilemma of many of the underdeveloped areas in Asia at the present. Many western writers have assumed that an increase in farm size is a prerequisite for increasing productivity; and the example of England in the eighteenth and nineteenth centuries is often cited as evidence. But Asia today has at least some parallels with early nineteenth-century England. Land is in short supply, labour relatively plentiful. The main aim in England then was to increase output per acre, and this is still the most necessary short-term aim for Asian countries. It must be doubted then if the amalgamation of small farms is necessary, yet this has often been the first aim of land improvement schemes. It is undeniable that the consolidation of fragmented holdings—perhaps the greatest benefit of enclosure in England—will help raise productivity, but it is less clear that the elimination of small farms is necessary in order to increase output per acre. The experience in South Lincolnshire suggests that this is not so; and this is confirmed

[1] G. E. Mingay, *Economic History Review*, XIV.

Some Conclusions

by events in Japan between 1880 and 1920. In this period there was no substantial alteration in farm size, yet crop yields rose by nearly 50 per cent. As in early nineteenth-century England the improvements introduced could easily be adopted by small farmers and increased output per acre rather than output per man.[1]

It was once thought that Parliamentary enclosure was responsible for the decline of the occupier owner in England. The use of the Land Tax returns to trace the fortunes of the small landowner has long since dispelled this view. In South Lincolnshire the situation was much the same as elsewhere in England. The occupier owner had been eliminated at some earlier date. The wars saw a slight increase in the numbers, just as the post-Napoleonic depression saw some reduction. But these changes were too slight to have had any impact on the rate of technical diffusion. More interesting is the fact that occupier owners in South Lincolnshire were concentrated into two regions, west Kesteven and the fenland. It may be that their importance here had an effect on the rate of technical progress, for it is often argued that occupier owners lacked the capital to make improvements, in contrast with the tenant who had the help of a landlord. This is perhaps a dubious assumption in itself; nor can it be held responsible for the slow progress of west Kesteven, for the effect of landownership cannot be distinguished from the other characteristics of the region—such as soil type—which helped retard improvement.

A further aspect of landownership—the size of great estates—has perhaps received scant attention in this study. Early writers believed that the great landowners of the eighteenth and nineteenth centuries were of great importance in furthering technical progress. The limited evidence on the size of estates suggests that the great estates of South Lincolnshire did not expand in the period under view. Nor is there much evidence that the landlords of the region were particularly active in promoting innovations. Before 1815 they were most certainly not, if Thomas Stone's strictures are to be believed. After 1815 many landlords helped their tenants underdrain and in some cases subsidized the use of oil-cake. But for the most

[1] B. F. Johnstone, 'Agricultural productivity and economic development in Japan', *Journal of Political Economy*, LIX, no. 6 (Chicago, Illinois, 1951), 500.

part the initiative towards improvements seems to have come mainly from the tenant farmers; this in fact accords with recent views on the role of great landlords in other parts of England.[1] Consequently it seems unlikely that the predominance of large estates on the Heath and the more divided nature of property in west Kesteven and the fenland had any effect on the regional development of agriculture in South Lincolnshire.

It will perhaps be many years before the true nature of the agricultural revolution in England is properly understood. The relationships between agricultural and industrial development in the eighteenth century are still too little known: the role of landlords and the nature of investment await further investigation. If, however, the experience of South Lincolnshire is to be a guide, any study of the impact of economic and technological change on agriculture must be approached from a regional standpoint. This is not simply a matter of differences in land use which arise to such a great extent from differences in soil type; the agricultural regions of South Lincolnshire were far more complex, for they varied not only in soil type and land use, but in farm-size, structure and landownership, accessibility and rentals, farming methods and productivity. Necessarily they responded differently to external changes in price and technique. Those who consider only the aggregate view of agricultural development sometimes forget that national change is no more than the sum of a number of regional changes; and before the whole can be fully comprehended, the parts must be analysed.

[1] F. M. L. Thompson, 'English great estates in the nineteenth century, 1790–1914', *Contributions to the First International Conference of Economic History* (Stockholm, 1960), pp. 385–97.

BIBLIOGRAPHY

Adams, L. P. *Agricultural Depression and Farm Relief in England, 1813–52*. London, 1932.
Adkin, B. W. *Land Drainage in Britain*. London, 1933.
Allen, G. R. 'Wheat farmers and falling prices', *The Farm Economist*, VII, no. 8. Oxford, 1954.
Anon. *The Effect of Free Trade on the Various Classes of Society*. Spalding, 1849.
Ashton, T. S. *An Economic History of England; the Eighteenth Century*. London, 1955.
Benedict, E. T., Stippler, H. H. and Benedict, M. R. *Theodore Brinkmann's Economics of the Farm Business*. Berkeley, California, 1935.
Bennet, M. K. 'British wheat yields per acre for seven centuries', III, no. 10, *Economic History*. London, 1935.
Board of Agriculture. *Communications to the Board of Agriculture*, IV. London, 1805.
Board of Agriculture. *The Agricultural State of the Kingdom in 1816*. London, 1816.
Brears, C. *Lincolnshire in the 17th and 18th Centuries*. London, 1940.
Burton, G. *Chronology of Stamford*. London, 1846.
Caird, James. *English Agriculture in 1850–51*. London, 1852.
Chambers, J. D. *Nottinghamshire in the Eighteenth Century*. London, 1932.
Chambers, J. D. 'Enclosure and the small landowner', *Economic History Review*, X, no. ii. London, 1940.
Chambers, J. D. 'Enclosure and the small landowner in Lindsey', no. 1, *The Lincolnshire Historian*. Lincoln, 1947.
Chambers, J. D. 'Enclosure and labour supply in the industrial revolution', *Economic History Review*, V, no. iii. London, 1953.
Clapham, Sir John. *An Economic History of Modern Britain*, I. Cambridge, 1926.
Clayton, K. M. 'The differentiation of the glacial drifts of the East Midlands', *The East Midland Geographer*, no. 7. Nottingham, 1957.
Collins, G. 'Cattle of the county', *The Lincolnshire Magazine*, I, no. 9. Lincoln, 1934.
Collins, G. 'Cattle of the county', *The Lincolnshire Magazine*, VII, no. 1. Lincoln, 1939.
Cossons, A. 'The turnpike roads of Nottinghamshire', *Historical Association Leaflet*, no. 97. London, 1934.
Cragg, W. *A History of Threekingham with Stow, in Lincolnshire*. Sleaford, 1913.

Bibliography

Creasey, John. *Sketches Illustrative of the Topography and History of New and Old Sleaford.* Sleaford, 1825.
Curtler, W. H. R. *The Enclosure and Redistribution of our Land.* Oxford, 1920.
Darby, H. C. *The Draining of the Fens.* Cambridge, 1940.
Darby, H. C. 'Some early ideas on the agricultural regions of England', *Agricultural History Review*, II. Oxford, 1954.
Davies, E. 'The small landowner, 1780–1832, in the light of the Land Tax assessments', I, no. i, *Economic History Review*. London, 1927.
Denman, D. R. and Stewart, V. F. *Farm Rents.* London, 1959.
Dietz, F. C. *An Economic History of England.* New York, 1942.
Dodd, A. H. *The Industrial Revolution in North Wales.* Cardiff, 1933.
Drescher, L. 'The development of agricultural production in Great Britain and Ireland from the early nineteenth century', *The Manchester School of Economic and Social Studies*, XXIII, no. 2. Manchester, 1955.
Ernle, Lord. *English Farming Past and Present.* London, 1932.
Fowler, G. 'Shrinkage of the peat-covered fenlands', *Geographical Journal*, LXXXI, no. 2. London, 1933.
Fuller, G. J. 'Development of drainage, agriculture and settlement in the fens of South East Lincolnshire, during the nineteenth century', *East Midland Geographer*, no. 7. Nottingham, 1957.
Fussell, G. E. 'Agricultural depression a century ago', *Journal of the Land Agents Society.* London, 1938.
Fussell, G. E. 'Steam cultivation in England' *Engineering.* London, 30 July 1943.
Fussell, G. E. *The Farmer's Tools.* London, 1952.
Fussell, G. E. 'The history of Cole', *Nature*, CLXXVI. London, 1955.
Fussell, G. E. and Compton, M. 'Agricultural adjustments after the Napoleonic Wars', *Economic History*, III, no. 14. London, 1939.
Glyn, J. 'Draining by steam power', *Transactions of the Royal Society of Arts*, LI. London, 1837.
Godwin, H. and Clifford, M. 'Studies in the post-glacial history of British vegetation', *Transactions of the Royal Society: Philosophical Transactions*, B, CCXX. London, 1938.
Gonner, E. C. K. *Common Land and Enclosure.* London, 1912.
Gray, H. L. 'Yeoman farming in Oxfordshire from the sixteenth century to the nineteenth century,' *Quarterly Journal of Economics*, XXIV. Boston, Mass., 1910.
Grigg, D. B. 'The 1801 crop returns for South Lincolnshire', *East Midland Geographer*, no. 16. Nottingham, 1961.
Grigg, D. B. 'The development of tenant right in South Lincolnshire', *The Lincolnshire Historian*, II, no. 9. London, 1962.

Bibliography

Grigg, D. B. 'Changing regional values during the Agricultural Revolution in South Lincolnshire', *Transactions and Papers of the Institute of British Geographers*, no. 30. London, 1962.

Grigg, D. B. 'Small and large farms in England and Wales; their size and distribution', *Geography*, XLVIII, no. 220. Sheffield, 1963.

Grigg, D. B. 'The Land Tax returns', *Agricultural History Review*, XI, part ii. Oxford, 1963.

Grigg, D. B. 'An index of regional change in English farming', *Transactions and Papers of the Institute of British Geographers*, no. 36. London, 1965.

Grigg, D. B. 'A note on rent in nineteenth century England', *Agricultural History*, vol. 39. Davis, California, July 1965.

Habbakuk, H. 'English landownership, 1680–1740', *Economic History Review*, XI, no. i. London, 1940.

Hallam, H. E. 'The new lands of Elloe', *Leicester University Department of Local History, Occasional Papers*, no. 6. Leicester, 1954.

Hasbach, W. *A History of the English Agricultural Labourer*. London, 1908.

Hatfield, Miss S. *The Terra Incognita of Lincolnshire*. London, 1816.

Havinden, M. 'Agricultural progress in open field Oxfordshire', *Agricultural History Review*, IX, part ii. Oxford, 1961.

Henderson, H. C. K. 'Agriculture in England and Wales in 1801', *Geographical Journal*, CXVIII, no. 3. London, 1952.

Hope-Jones, A. *The Income Tax in the Napoleonic Wars*. Cambridge, 1939.

Hosford, W. H. 'The enclosure of Sleaford', *Lincolnshire Architectural and Archaeological Society, Reports and Papers*, VII, no. 1. Lincoln, 1957.

Hoskyns, W. G. 'Regional farming in England', *Agricultural History Review*, II. Oxford, 1954.

Hunt, H. G. 'The chronology of Parliamentary enclosure in Leicestershire', *Economic History Review*, X, no. ii. London, 1957.

Hutchinson, Samuel. *Practical Instructions on the Draining of Land*. Grantham, 1847.

Jackman, W. T. *The Development of Transportation in Modern England*, I. London, 1916.

Jacks, G. V. *Soil*. London, 1954.

James, P. E. and Jones, C. F. *American Geography; Inventory and Prospect*. Syracuse, New York, 1954.

John, A. H. 'The course of agricultural change, 1660–1760', in *Studies in the Industrial Revolution*, ed. L. S. Pressnell. London, 1960.

Johnson, A. F. *The Disappearance of the Small Landowner*. London, 1909.

Bibliography

Johnstone, B. F. 'Agricultural productivity and economic development in Japan', *Journal of Political Economy*, LIX, no. 6. Chicago, Illinois, 1951.

Jones, E. L. 'The changing basis of English agricultural prosperity, 1853-73', *Agricultural History Review*, X, part ii. Oxford, 1962.

Kerridge, E. 'Turnip husbandry in High Suffolk', *Economic History Review*, III, no. iii. London, 1956.

Lavergne, L. de. *The Rural Economy of England, Scotland and Ireland*. Edinburgh, 1855.

Layton, W. T. 'Wheat prices and the world's production', *Journal of the Royal Statistical Society*, LXX. London, 1909.

Levy, H. *Large and Small Holdings*. Cambridge, 1911.

The Lincolnshire Cabinet and Intelligencer. Lincoln, 1827.

Linton, D. L. 'The landforms of Lincolnshire', *Geography*, XXXIX, no. 2. Sheffield, 1954.

Marrat, W. *A History of Lincolnshire*. Boston, 1814-16. Boston, four volumes.

Marshall, C. E. *Guide to the Geology of the East Midlands*. Nottingham, 1948.

Marshall, T. H. 'Jethro Tull and the New Husbandry of the eighteenth century', *Economic History Review*, II, no. i. London, 1929.

Mills, D. R. 'The regions of Kesteven devised for the purpose of agricultural history', *Reports and Papers of the Lincolnshire Architectural and Archaeological Society*, VII, part 1. Lincoln, 1957.

Mills, D. R. 'Enclosure in Kesteven', *Agricultural History Review*, VII, part ii. Oxford, 1959.

Minchinton, W. E. 'Agricultural returns and the government during the Napoleonic Wars', *Agricultural History Review*, I. Oxford, 1954.

Mingay, G. E. 'The agricultural depression, 1730-1750', *Economic History Review*, VIII, no. iii. London, 1956.

Mingay, G. E. 'The size of farms in the eighteenth century', *Economic History Review*, XIV, no. iii. London, 1962.

Mingay, G. E. *English Landed Society in the Eighteenth Century*. London, 1963.

Mingay, G. E. 'The agricultural revolution in English history: a reconsideration', *Agricultural History*, XXXVII, no. 3. Urbana, Illinois, 1963.

Mitchell, J. B. (Ed.). *Great Britain: Geographical Essays*. Cambridge, 1962.

Nicholson, H. N. *The Principles of Field Drainage*. Cambridge, 1942.

Orwin, C. S. 'The History of tenant right'. *Agricultural Progress*, XV. London, 1938.

Padley, J. S. *Fens and Floods of Mid-Lincolnshire*. Lincoln, 1882.

Bibliography

Parker, R. 'Coke of Norfolk and the Agricultural Revolution', *Economic History Review*, VIII, no. iii. London, 1955.

Plumb, J. H. 'Sir Robert Walpole and Norfolk Husbandry', *Economic History Review*, V, no. i. London, 1952.

Porter, G. R. *The Progress of the Nation*. London, 1847.

Porter, H. 'Old Private Banks of Southern Lincolnshire', *The Lincolnshire Magazine*, III, no. 2. Lincoln, 1935.

Porter, H. 'Lincolnshire private bankers', *The Lincolnshire Magazine*, V, no. iii. Lincoln, 1937.

Prentice, J. E. 'The subsurface geology of the Lincolnshire fenland', *Transactions of the Lincolnshire Naturalists Union*, 12. Louth, 1950.

Pressnell, L. S. *Country Banking in the Industrial Revolution*. Oxford, 1956.

Prince, H. 'The tithe surveys of the mid-nineteenth century', *Agricultural History Review*, VII. Oxford, 1959.

Riches, N. *The Agricultural Revolution in Norfolk*. Chapel Hill, North Carolina, 1937.

Ross, E. D. and Toritz, R. L. 'The term "Agricultural Revolution" as used by agricultural historians', *Agricultural History*, XXII, no. 1. Washington, 1948.

Russell, E. J. *Soil Conditions and Plant Growth*. London, 1956.

Sidney, Samuel. *Railways and Agriculture in North Lincolnshire*. London, 1848.

Skertchley, S. J. K. *The Geology of the Fenland*. London, 1877.

Smith, G. *The Land of Britain*, part 69. London, 1937.

Stamp, L. D. *The Land of Britain*, parts 76 and 77. London, 1942.

Stone, Thomas, *A General View of the Agriculture of the County of Lincoln*. London, 1794.

Stone, Thomas, *A Review of the Corrected Agricultural Survey of Lincolnshire by Arthur Young*. London, 1800.

Swinnerton, H. H. and Kent, P. E. *The Geology of Lincolnshire*. Lincoln, 1949.

Tate, W. E. 'The cost of Parliamentary enclosure', *Economic History Review*, V, no. ii. London, 1952.

Thirsk, J. *English Peasant Farming*. London, 1957.

Thirsk, J. and Imray, V. (ed.). 'Suffolk farming in the nineteenth century', *Suffolk Records Society*, no. 1. Ipswich, 1958.

Thomas, D. 'The acreage returns of 1801 for the Welsh borderland', *Transactions and Papers of the Institute of British Geographers*, no. 26. London, 1959.

Thompson, F. M. L. 'The English land market in the nineteenth century', *Oxford Economic Papers*, new series, IX, no. 3. Oxford, 1957.

Bibliography

Thompson, F. M. L. 'English great estates in the nineteenth century, 1790–1914,' *Contributions to the First International Conference of Economic History*. Stockholm, 1960.

Thompson, F. M. L. *English Landed Society in the Nineteenth Century*. London, 1963.

Thompson, Pyshey. *Collections for a Topographical and Historical Account of Boston and the Hundred of Skirbeck*. London, 1820.

Thompson, Pyshey. *The History and Antiquities of Boston*. Boston, 1856.

Thompson, R. J. 'An inquiry into rent in the nineteenth century', *Journal of the Royal Statistical Society*, LXX. London, 1907.

Tomes, F. M. 'The draining of the Witham Fens', *The Lincolnshire Magazine*, II. Lincoln, 1934.

Tritton, W. 'The origin of the threshing machine', *The Lincolnshire Magazine*, XI, no. 2. Lincoln, 1943.

Trueman, A. E. 'The Lias brickyards of west Kesteven', *Transactions of the Lincolnshire Naturalists Union*, IV. Louth, 1916–18.

Turnor, E. *Collections for a History of the Town and Soke of Grantham*. London, 1806.

Venn, J. A. *The Foundations of Agricultural Economics*. London, 1933.

The Victoria History of the County of Lincoln, II. London, 1906.

Weaver, J. C. 'Changing patterns of cropland use in the Middle West', *Economic Geography*, XXX, no. 1. Worcester, Mass., 1954.

Wheeler, W. H. *History of the Fens of South Lincolnshire*. Boston, 1896.

White, C. *A Short and Plain Letter on Agricultural Depression*. London, 1816.

Whittlesey, D. 'Major agricultural regions of the earth', *Annals of the Association of American Geographers*, XXVI. Lancaster, Pennsylvania, 1936.

Woodruffe Peacock, A. *Fenland Soils*. Spalding, 1906.

Young, Arthur. *A General View of the Agriculture of the County of Lincoln*. London, 1799.

Young, Arthur. *A Farmer's Tour through the East of England*, I. London, 1771.

Youngson, A. *Possibilities of Economic Progress*. Cambridge, 1959.

INDEX

Accessibility, and grazing, 69
Acts of Parliament, 27, 28, 35, 39, 46
Adams, L. P., 117 n., 119 n.
Adkin, B., 141 n.
Adventurers, 30, 50
Agricultural Area,
  increase in England, 3
  increase in South Lincolnshire, 70–1
  maximum, 62
Agricultural
  depression, 38, 117–25, 129, 130, 131, 173
  output, statistics of, 33
  prices, 4, 9, 34, 37–8, 40, 52, 62, 66, 111, 117–22, 126–7, 128–9, 135–6, 142, 155, 156, 157, 159, 173–4, 180; *see also* Wool, Wheat, Beef and Mutton Prices.
  regions, delimitation of, 95–99; *see also* West Kesteven, North-west Kesteven, etc.
  rent, *see* Rent
  Revolution, 1, 2, 8, 9, 47, 190, 191, 192, 198
Algakirk, 164
  occupier owners in, 172
Allen, G. R., 159 n.
Amalgamation of farms, 166, 168
  see also Farm size
America, 150
American War, and wool prices, 38
Ancaster, 103, 122, 123, 124, 125
  Duke of, 83, 85, 90, 106; *see also* Lord Willoughby
  estate, 91, 128, 156; farm size on, 91
Ancaster Gap, 12, 13, 14, 15, 20, 104
Ancholme Valley, 13
Anwick, 44
Arable, 47, 50, 63, 64, 66, 76, 112, 157
  after drainage, 28
  arable grass ratio, 96
  expansion of, 70, 71, 126, 155
  farmers, in distress, 121, 158
  farming, new methods, 7
  increase in marsh, 71
  management of, 78
Artificial grasses, *see* Temporary grasses
Ashton, Professor T. S., 36 n., 89, 89 n.

Asia, farm size in, 198
Aslackby, 143
Australia, export of rams to, 150
Average size, of farm, 91
  see also Farm size

Banks
  country, 5, 40, 121–2
  jointstock, 122
Bardney, 28, 138
Bare fallow, 56, 108, 110
  see also Fallow
Barley, 17, 19, 49, 74, 96, 104, 111, 158, 162, 183, 185
  prices, 120
  yields, 55
  see also Malting barley, Maltsters
Bassingthorpe, 105
Beans, 19, 55, 72, 73, 76, 77, 96, 106, 110, 147, 163, 164
Beckingham, 121
Beef
  cattle, 79, 110, 118, 120, 183
  prices, 118, 120
  see also Cattle, Breeding
Belvoir, Vale of, 69
Benedict, E. T., Benedict, M. R. and Stippler, H. H., 186 n.
Bennet, M. K., 2 n., 59 n.
Bentall's Plough, 150
Bevan, B., 31 n.
Billingborough, 121
Billinghay, 44, 88, 147
Black Sluice, the, 28, 29, 140
  Act, 35
  Level, 26, 29, 138, 139
Blankney, 83
Blue Buttery Clay, 142, 149
Blundell, Phillip, 83
Board of Agriculture
  Circular Inquiry, 120
  returns, 16, 95, 164, 165
Bones, 3, 4, 49, 56, 57, 134, 148, 162
Boston, 25, 26, 28, 29, 40, 42, 43, 45, 46, 68, 70, 75, 121, 138, 139, 150, 157, 158, 184
  exports of grain from, 71, 162
Boston Haven, 27, 28, 29, 138

207

Index

Boston, Lord, 36, 83
Boulder Clay, 106, 111, 142, 147
Bourne, 28, 29, 35, 40, 43, 44, 139
Bourne Eau, 35, 44
Bower, A., 28 n.
Bradford, 120
Branston, 121
Brant, river, 15, 100
Brayford Head, 100
Brears, C., 43 n.
Breeding, of livestock, 53, 57, 106, 110
Breweries, 100, 111
Brinkmann, Theodore, 186
Bristol, Earl of, 36, 37, 67, 83
Brothertoft, 78
Brownlow, Lord, 64, 83, 143
Burton, G., 38 n.
Burton Pedwardine, 44

Cabbages, 183
Caird, James, 126, 128, 128 n., 129 n., 134, 134 n., 135 n., 137, 137 n., 142, 142 n., 147, 163 n., 193 n.
Calcraft, J., 122
Calthorpe, George, 128
Cambridge colleges, as landlords, 83
Cambridgeshire, 17, 149, 157
Canals, 4, 5, 35, 45
Canwick, 36, 60
 rent in, 36
Capital, 5, 101, 128
 for Witham drainage, 26
 improvements, 39
 tenants paying rent from, 123
Car Dyke, 140
Carline, Richard, 135
Cartwright, John, 45, 46, 78
Castle Bytham, rent in, 36
Castlereagh, Lord, 121
Catch cropping, 67
Catchwater, 28
Cattle, 53, 57, 78, 80, 100, 106, 110, 118, 120, 166
 densities, 79, 163-4, 183-4
Cattle/sheep ratio, 165
 see Breeding, Livestock, Beef
Census, of 1851, 6, 168, 170
Chalky Boulder Clay, 15, 19, 20
Chambers, J. D., 5 n., 50 n., 89 n.
Channel Islands, cattle from, 57
Chapel Hill, 26, 27, 45
Chaplin, Charles, 83, 156
Chapman, W., 27 n.

Cheshire, rent per acre in, 194
Chicory, 164
Cholmley, Sir Montague, 83
Christ's Hospital, 83, 125, 133, 135, 166
Church estates, 83
Clapham, Sir John, 4 n., 117 n., 127, 127 n.
Clarges, Sir Thomas, 83
Clarke, John, 144, 148, 152, 160, 161, 164, 183
Clayhole, 27, 138
Claying of peat fens, 128, 132, 134, 145, 161, 185, 189
 cost of, 149
 in Cambridgeshire, 149
Claypole, amalgamation of farms in, 167
Clays, 15, 19, 20, 22, 47, 49, 76, 106, 111, 142, 147, 149
 costs of cultivation, 101
 of England, 194
 and grass, 68
 rotations on, 56
 soils on, 21
 undrained, 18, 101
 unsuitable for New Husbandry, 102, 178
 see also West Kesteven, South-east Kesteven, Underdrainage, Heavy soils
Clayton and Shuttleworth, 151
Cliff Row, 104
Clifford, M., 17 n.
Clover, 45, 48, 55
Coastal banks, 23, 31
Coffee, 164
Coke of Holkham, 190
Coleseed, 56, 67, 76, 163
 see also Rape
Collins, G., 79 n., 150 n.
Communications, improvements in and rent, 180
Compton, M., 117 n.
Consolidation of holdings, 1, 109, 198
Corby, 107
Cornbrash, 13
Corn Laws, 158
 and increase of arable, 119
 repeal of, 123
Cossons, A., 43 n.
Cottagers, 90, 197
Counterdrain, 31
Country banks, 5, 40, 121-2
County Rate, 176

Index

Court of Sewers, verdict of, 86
Covenants of improvement, 62, 133
Cowbitt Wash, 90, 161
Cragg, John, 45, 46 n., 49 n., 67 n., 69, 82, 167
Cragg, W., 80 n.
Cranwell, 83, 168
Creasey, J., 9 n., 35 n., 37 n., 44 n., 55 n., 57 n., 64 n., 67 n., 77 n., 79 n., 85 n., 122 n., 128 n., 142 n., 151 n.
Creasey, Richard, 122
Creech soils, 106, 111
Crimean War, 166, 168
   and agricultural prices, 119, 120
Crop combinations, 96, 104, 108
   and soil type, 111
Crop yields, 3, 18, 52, 58, 60, 61, 62, 100, 112, 148, 152-3, 159, 166, 185, 186, 191
   on clays, 19
   and farm size, 94
   in north-west Kesteven, 103
   and soil fertility, 188
   see also Wheat yields, Barley yields
Crops, *see under* individual crops: Wheat, etc.
Cross Drain, 31
Cross Keys Wash, 141
Crosskill's Clod Crusher, 150
Crown estates, 83
Cubitt, W., 140 n.
Cuesta, of Lincolnshire Limestone, 13, 14
Curtler, W. H. R., 1 n.
Custom of the County, 134
   see also Tenant right

Dairying, 79
Darby, H. C., 23 n., 95 n.
Davies, E., 1 n., 85 n.
Declining crop yields, 64
   see also Soil exhaustion
Deeping Fen, 21, 23, 26, 30, 31, 40, 50, 53, 54, 67, 70, 87, 107, 108, 140, 144, 147, 148
   grazing rights in, 52
   intercommoning in, 107
Deficit financing, of estates, 132
Denman, D. R., 99 n.
Dietz, F. C., 193 n.
Digby, 128
Dip slope, of Lincolnshire Limestone, 13
Dodd, A. H., 3 n.

Doddington, 37, 62, 128, 142
   farm size in, 168
   Moor, 134
Dorrington Fen, claying of, 149
Dorset, 48
Drainage
   Districts, 26
   of clay soils: in south-east Kesteven, 144; in South Holland, 144; *see* Underdrainage
   of Fens, 23-32, 137-44; of East, West and Wildmore Fens, 70; cost of, 39; rate, 39; *see also* Drains, Outfalls, Scoopwheels, Flooding, Silting, Windmills, Enclosure
   of light lands, 19
Drains
   Main, 32, 137, 141
   major, 25, 28
   minor, 25, 27
   as navigable waterways, 45
   parish, 23
Drescher, L., 3 n.
Drill, 2, 4, 48, 53, 150, 192
Dry valleys, 13

East Anglia, 129, 154
   stall feeding in, 146
   turnips in, 190
East Fen, 21, 39, 52, 54, 67, 70, 78, 138, 142, 144
East Kesteven, 18, 98
East, West and Wildmore Fens, 17, 26, 28
Eastern England, 145, 153, 192, 193, 194, 195, 196, 197
Ebb tide, 25, 26, 27
Edenham, 105, 151
Edwards, K. C., 13 n., 19 n., 20 n.
Ellison, Richard, 44
Enclosure, 1, 5, 6, 8, 22, 27, 30, 33, 45-53, 55, 82, 104, 107, 108, 112, 126, 132, 153, 169, 188, 189, 199
   and cottagers, 90, 197
   of East, West and Wildmore Fens, 28, 52
   financing of, 39
   of Holbeach, 23
   and livestock, 113
   of marsh, 109
   and occupier owners, 87
   and small farmers, 89-90
   see also Old Enclosure
Engrossing of farms, 1, 6, 89
   see also Amalgamation and farm size

Index

Ernle Lord, 117 n.
Estates, 90, 91, 199–200, 132
  accounts, 123–5
  deficit financing of, 132
  surveys of, 89
Ewerby, 105
Ewerby Thorpe, 44
Expansion of arable acreage, 70–1, 126, 155
Exports, of grain from Boston, 71, 162
Extensive farming, 113

Fallow, 18, 47, 72, 100, 108, 110
  on clays, 56
  in open fields, 53
  see also Bare fallow
Farm machinery, 1, 5, 9
  clod crusher, 150
  drill, 2, 4, 48, 53, 150, 192
  harrow, 151
  hoe, 2, 4, 53
  mole plough, 143
  reaper, 1, 4, 150
  steam ploughing, 151
  thresher, 151
Farm size, 1, 6, 7, 88–94, 107, 109, 112, 113, 166–71, 191, 197, 198
  and relief, 106
  average, 92
  see also Engrossing, Amalgamation, Fragmentation, Gavelkind, Enclosure
Farms, unlet, 121
Fascine training, 138
Fenland, the
  claying in, 149
  crop yields in, 59–61
  drainage of, 22–32, 137–41, 175
  enclosure in, 51–3, 54
  farming of, 107–10, 147
  farm size in, 92, 93, 167, 170–1
  increase in, arable in, 157, 159, 161
  land use in, 67–8, 70, 107, 164
  occupier owners, 173
  origins and relief, 17–18
  roads in, 43
  soils, 20–2, 64, 178, 179
  see also Interior Fens, the Marsh, the Townlands, South Holland, Drainage of the Fens, Flooding in the Fens, Kirton, Skirbeck, Deeping Fen
Fertilizers, 2, 3, 4, 19, 46, 49, 56, 57, 112, 126, 132, 134, 138, 153, 162, 185, 198
  see also Bones, Guano, Lime

Flax, 27
Flooding
  of Cowbitt Wash, 161
  in fens, 24, 25, 26, 27, 29, 30, 31, 32, 43, 67, 109, 111, 139, 141, 161
  of Marsh, 109
  of Witham, 27, 100
Fodder crops, 74, 79, 145, 163, 182
Folding of sheep, 18, 20, 64, 102, 105, 166
  see also Sheep
Folkingham, 68
Foot rot, in sheep, 64, 100, 103, 179
Fortescue, Earl, 83, 121
Fosdyke, 172
Fosdyke Bridge, 31, 39, 42, 140
Foss dyke, 35, 44, 45, 46
Fowler, G., 21 n., 45 n.
Fowler's ploughing set, 151
Fragmentation of farms, 147, 169
Free Trade, 128
Freeholders, 52, 167
  see also Occupier owners
Fuller, G. J., 110 n.
Fussell, G. E., 76 n., 117 n., 141 n., 151 n.

Gangs, agricultural, 4, 192
Gavelkind, 93, 94, 113
Geology, of South Lincolnshire, 13, 18
George III, 13
Gilbert, Mr, tenant of Lord Willoughby, 146
Glacial deposits, 15
Glen, river, 24, 25, 29, 30, 31, 35, 44, 53, 139, 140
Glyn, J., 140 n.
Godwin, H., 17 n.
Golding, 29 n.
Gonner, E. K. C., 1 n.
Gorse, 66, 103, 156
Gosberton, 84
Graffoe Hills, 16
Grain, 32, 182
  acreage in 1801, 71, 73
  exports of, 71
  prices of, 11, 52, 62, 66, 157
Grand Sluice, 26, 27, 28, 79, 138, 139
Grantham, 40, 44, 46, 50, 52, 54, 73, 83, 91, 100, 101, 103, 105, 121, 131, 146, 150, 151, 171, 172
Grantham Canal, 35, 39, 44, 45, 46
Grantham–Bridgend turnpike, 42
Grassland, permanent, 3, 18, 19, 51, 54, 112, 117, 162, 166, 180–1, 189

210

Index

conversion to arable, 155-7, 159, 160
and delimitation of agricultural regions, 96-7
and farming innovations, 63
and livestock densities, 78-80
and mixed farming, 126, 160
and old enclosure, 50, 113
and rent, 101
and soil type, 68-9, 113
in south-east Kesteven, 106
and tithes, 69
in the Townlands, 109-10
in west Kesteven, 100-1
Gravels, 17, 18, 19, 20, 52, 66, 107
Gray, H. L., 85 n.
Great Hale, 29
Great North Road, 41
Grigg, D. B., 37 n., 72 n., 87 n., 96 n., 99 n., 123 n., 133 n., 170 n., 193 n.
Grimsthorpe, 83
Guano, 3, 134, 148
Gunthorpe Sluice, 31, 141
Gutherham Gowt, 28
Guy's Hospital, 83

Habbakuk, H., 5 n., 132, 132 n.
Hale Fen, 29
Hammond Beck, 139
Hampshire, 48
Hare, Mr, 29 n.
Harrow, 151
Harts Ground, 45
Hasbach, W., 4 n., 127 n.
Hatfield, Miss S., 67 n.
Havinden, M., 2 n.
Hawkes, Thomas, 90
Heath, the, 16, 22, 37, 40, 50, 53, 56, 59, 64, 65, 77, 79, 80, 81, 88, 94, 99, 101, 102, 106, 108, 109, 111, 112, 113, 121, 127, 147, 155, 161, 163, 164, 165, 172, 177, 179, 181, 183, 189, 193, 200
  absence of occupier owners, 85
  crops in 1801, 73-6
  crop yields, 60, 152-3, 185-6
  drill, 150
  farms, size of, 91-3, 171
  fertilizers, 57, 148
  improvements after enclosure, 55
  increase in arable, 70, 156
  increases in rent, 178
  mixed farming on, 145
  open fields on, 52
  relief, 13-15
  rotations, 146
  soils, 19-20
  type of farming, 103-5, 162
  wastes on, 66
Heathcote, Sir Gilbert, 36, 83, 85, 106
Heavy soils, 3, 7, 8, 18, 69, 110, 113, 155, 178, 179, 193
Heckington, 20, 29, 143
Hemp, 77
Henderson, H. C. K., 72 n., 77 n.
'High' farming, 2, 9, 126, 146, 194, 196
High Feeding, 146
Hobhole, 25, 28, 138
Hodson, Mr, tenant of Lord Willoughby, 135
Hoe, 2, 4, 53
Holbeach, 23, 121, 164, 170
Holland, 17, 21, 29, 41, 50, 52, 54, 60, 68, 70, 75, 77, 84, 85, 86, 87, 88, 91, 93, 94, 98, 107, 164, 166, 172, 176
Holland Fen, 26, 27, 28, 39, 43, 75, 138, 147
  rights of common in, 52
Home farms, 82
Hope-Jones, A., 102 n.
Horbling, 121
Hornsby, Richard, 151
Horses, 74
Hosford, W. H., 90 n.
Hoskins, W. G., 95 n.
Hough on the Hill, 14
Hougham, 168
Hume, Sir A., 84
Humus content of soils, 20, 22
Hunt, H. G., 50 n.
Huntingdon, 157
Hutchinson, Samuel, 143, 143 n.

Improvements in farming methods, 1-3, 7-8, 33-5, 41-6, 142, 143, 148
  on clay soils, 18-19, 47-9
  and crop yields, 152-3
  effect on crop selection, 162-3
  and enclosure, 55
  and farm machinery, 150-1
  and farm size, 198
  and labour needs, 59, 128
  and landlords, 63-4, 82, 131-2
  and livestock, 149-50
  and mixed farming, 58, 144-5
  and the Norfolk four-course, 76
  in open fields, 53

Index

Improvements in farming methods (*cont.*)
  rate of improvement, 102, 104, 105, 107, 113–14, 189–92
  regional differences, 64–5, 175–80
  rotations, 146
  stimulated by low prices, 126–9
  and tenants, 62–3
  and tenant right, 132–3, 135
  underdrainage, 142–4
Income Tax, 177, 194
Industrial cities, as markets, 46
Inflation, and country banks, 40
Inheritance, of farms in the fenland, 167
Intercommoning, 54
  in Deeping Fen, 107
  in East, West and Wildmore Fens, 17, 28
Interior Fens, the, 56
  claying in, 149
  conversion of arable, 70, 157
  definition of, 22
  drainage of, 24–32, 137–41
  effect of depression on, 123
  enclosure in, 50
  farm size in, 171
  fall on crop yields, 60
  flooding in and roads, 42
  increase in rent, 178–9
  land use in, 180–1
  oats in, 73
  paring and burning in, 86
  small farmers in, 90
  summer grazing in, 54, 81, 184
  turnips in, 175
  type of farming in, 110–11
  underdrainage in, 144
Interest rates, 173, 174
Investment, 51, 188, 200
  in depression, 130–2
  in eastern England, 196
  landlord, cf. tenant, 41
  provision of capital by banks, 39–40
  reluctance of Lincolnshire landlords, 63, 65
Ireland, 57
Irish
  cattle, 80
  labour, 41, 192
Ironbridge, 138

Jackman, W. T., 43 n.
Jacks, G. V., 19 n.

James, P. E., 96 n.
Japan, farm size and crop yields in, 198, 199
Jarvis, Colonel, 37, 62, 119
Jarvis estate, the, 142
Jessop, W., 27 n.
John, A. H., 193, 193 n.
Johnson, A. H., 89 n.
Johnstone, B. F., 199
Jointstock banks, 122
Jones, C. F., 96 n.
Jones, E. L., 145 n.
Jurassic limestone, 17

Kent, Nathaniel, 58, 69
Kent, P. E., 13 n.
Kerridge, E., 48 n., 190 n.
Kesteven, 15, 17, 18, 21, 22, 39, 52, 57, 60, 73, 75, 78, 84, 93, 141, 167
  acreage and output of grains, 70–1
  County Rate in, 176
  crop yields in, 59, 179
  Land Tax returns for, 85 ff., 172
  landlords in, 83
  livestock in, 79
  open fields in, 50
  riots in, 150
  see also North-west Kesteven, West Kesteven, South-east Kesteven, North-east Kesteven, Kesteven Fenland
Kesteven Fenland, 17, 21, 98, 107, 108, 121, 138, 147, 172
Keyworth, 45
Kinderley's Cut, 32
King, Neville, 83
King Fane, W. V. R., 49 n.
Kings Lynn, 43
Kingston, Duke of, 129
Kirton, 52, 53, 86, 93, 108, 109, 111
Kirton Hundred, 23, 52, 92, 99, 143, 144, 148

Labour, 58, 192
  in arable and grazing, 4
  for claying, 149
  costs, 40, 127
  increase of, in New Husbandry, 3, 48
  for industrial towns, 5
  and new methods, 128
  riots, 150
  troubles on Witham, 27
Lancashire, 46, 77, 194

212

Index

Land Tax returns, 6, 9, 82, 85, 86, 87, 88, 89, 91, 172, 173, 199
Land use, 45, 47
  changes after wars, 117
  in common fen, 52
  declining regional differentiation, 180–3
  delimitation of agricultural regions, 95–8, 111–12
  in Fenland, 107–11
  and flooding of fens, 27, 28, 29, 54, 70, 75, 76, 81, 110, 111, 140–1
  on Heath, 103–5
  impact of low prices on, 155–9
  improving methods and, 161–4
  landlord policy on, 69
  on Marsh, 164
  and mixed farming, 48–9, 126, 144–9, 150–60
  in Napoleonic Wars, 66–81
  and 'old' enclosure, 50
  and open fields, 76
  and regional pattern of, 160–1
  and rent, 179–182
  and soil type, 18–19, 21, 66, 101–2, 103, 104, 106, 112–14
  in south-east Kesteven, 105–7
  and transport, 45, 69, 107
  in west Kesteven, 100–2
Landlords, 121, 189
  absence of large in Holland, 83–4
  and cattle breeding, 49, 57
  and cost of improvement, 1, 5, 39–41
  deficit financing, 132
  in depression, 129
  and enclosure, 52
  and expenditure, 130–2
  farm size in estates, 90–1
  lack of enterprise, 63
  and policy on grass, 69
  and policy on rental changes, 36–7, 118–19, 125
  and rents, 122–5
  residences of, 83
  role in improvement, 82–3, 199–200
  and tenant right, 132–6
  and underdrainage, 142–3
  and wheat prices, 118–19
Landownership, 1, 5, 6
  and agricultural regions, 99
  changes in after 1815, 167, 172–4
  changes in during Napoleonic Wars, 85–8
  and farm size, 93–4

  in Fenland, 107
  on the Heath, 105
  in north-west Kesteven, 103
  pattern of, 83–4
  and regional differences in improvements, 65
  and soil type, 112–13
  in south-east Kesteven, 106
  in west Kesteven, 101
Large farms, 64, 65, 92, 94, 103, 105, 168, 170, 171, 175
  see also Amalgamation, Engrossing, Farm size
Laughton, 90
Lavergne, L. de, 137 n.
Layton, W. T., 119 n.
Leadenham, 43
Leaf, John, 167
Leases, 62, 133, 135, 193
Leicester sheep, 39
Leicestershire, 13, 14, 39, 80
Levy, H., 88, 89, 89 n., 90, 158, 166, 167 n.
Lewin, W., 140 n.
Lias Clay, 15, 16, 18, 19, 50, 98, 143
Light lands, 2, 6, 18, 19, 47, 64, 77, 135, 162, 178, 179, 183, 193
Lime, 3, 19, 56, 76, 100, 148
Lincoln, 13, 16, 17, 20, 26, 28, 40, 42, 46, 50, 66, 97, 100, 101, 103, 105, 138, 139, 147, 151, 171
  Corporation, 44
Lincoln Edge, 161
Lincoln Red, 57, 149
Lincolnshire
  limestone, 13, 19, 20, 88, 97
  longwool, 34, 39, 57, 120; see also sheep
Lindsey, 17, 28
Linton, D. L., 13 n.
Livestock, 175
  breeding of, 57, 149–50
  breeding and open fields, 49, 53
  and common fen, 52, 54
  farming, effect of prices on, 38, 120, 122, 158–9
  farming and old enclosure, 50
  in Fenland, 107–10
  and mixed farming, 48, 58, 64, 113–14, 126, 144–6, 159–60
  in north-west Kesteven, 103
  regional densities and type of farming, 164–6, 183–5

213

Index

Livestock (*cont.*)
  in south-east Kesteven, 106
  in west Kesteven, 100
London, 46, 193
Long Bennington, 50, 84, 100
Long Sutton, 164
Lord's Drain, 30, 31
Low prices, and improvement, 128–9
Lothians, the, 146
Low tide, 25, 32
Lower Welland, 39

MacAdam, 43
Machinery
  farm, *see* Farm machinery
  in agricultural revolution compared with industrial revolution, 4
Main drains, 32, 137, 141
Malting barley, 75, 100, 120, 179
Maltsters, 76, 100, 162
Manchester, 46
Mangolds, 183
Manners, Sir William, 83
Manufacturers, of farm implements, 151
Manure, 19, 56, 145, 162
Market Deeping, 17, 20, 30, 38, 52, 53, 76, 107, 111, 188
Market towns, 42, 46, 107, 108, 194
  effect on rent, 101
  sale of wheat in, 157, 158
Marling, 63
Marlstone, 13, 14, 18, 19, 20, 100
Marrat, W., 39 n., 57 n., 59, 60 n., 67, 68, 68 n., 70 n., 71 n., 78 n.
Marsh, the, 31, 56, 70, 75, 76, 98, 99, 110
  drainage of, 23
  farm size in, 171
  farming in, 109, 164
  grass in, 68
  increase of arable in, 71
  soil type, 21–2, 64
Marshall, C. E., 19 n.
Marston, 167
Maud Foster Drain, 25, 28
Maud Foster Sluice, 28
Maxwell, George, 31, 70, 137
Meat, 46
Midlands, 47, 153, 193, 194
Millers, 78
Mills, D. R., 16 n., 49 n., 103 n.
Minchinton, W. E., 72 n.
Mingay, G. E., 89 n., 129 n., 192 n.

Mitchell, J. B., 13 n.
Mixed farming, 2, 3, 48, 58, 114, 126, 144–5, 159, 166, 180, 183, 189, 194, 196
Mole ploughing, 143
Mortgages, 121
Morton, 29
Mossop, John, 74
Moulton, 36
Mustard seed, 78, 164
Mutton, 80, 150
  price of, 38, 120
Mylne, W. S., 140 n.

Nene, river, 24, 25, 31, 32, 43, 141
Nesbitt, John, 134
Ness, Hundred, 60
Neville, Rev. C., 133
New Domesday Book, 82
New Husbandry, 3, 47, 48, 58, 113, 190, 192
New Leicester, 57
New Zealand, 150
Newark, 41, 43, 46, 76, 100, 101, 111
Nicholson, H., 141 n.
Nitrogen content of soil, 3, 49, 60, 77, 112, 162, 185
Norfolk, 17, 47, 63, 80, 137, 157
  four-course, 2, 4, 48, 146, 148, 183
  landlords, 189
  leases, 62
  system, 58, 64, 113, 177, 178, 195
North Drove Drain, 31, 140
North-east Kesteven, 104
North Kyme, 44, 84
North Lincolnshire, 39, 149
  see also Lindsey
North Main Level Drain, 31
North Sea, 44
North-west Kesteven, 18, 83, 105, 146, 166, 175, 178, 189
  crops in, 73, 76, 98, 160, 162, 181, 182
  farming in, 102–3
  farm size in, 168, 170
  fertilizers, 148
  geology, 16
  increase in arable, 156
  livestock, 164, 165, 184
  rent in, 99, 179
  soils, 19–20, 64
  waste, 66, 156, 180
Northamptonshire, 80
Northumberland, 57, 154, 193

214

Index

Nottingham, 46
Nottinghamshire, 100
Nucleated villages, 22, 101

Oats, 19, 46, 59, 67, 70, 72, 73, 74, 77, 96, 103, 104, 108, 109, 110, 120, 147, 161, 162, 185, 188
Occupier owners, 1, 5, 6, 65, 84, 85, 86, 87, 93, 101, 103, 105, 106, 107, 113, 172, 173, 175, 199
Oil-cake, 49, 56, 134, 135, 146, 198, 199
'Old' enclosure, 50, 51, 100, 113
Old Shire Drain, 31
Open fields, 47, 52, 53, 66, 103, 191
Ordnance Survey maps, 22
Orwin, C. S., 133 n.
Osbournby, 90
Outfalls, of rivers, 25, 27, 32, 137, 138, 140, 141
Oxen, 74
Oxford Clay, 16, 18, 19
Oxford colleges, 83

Padley, J. S., 20 n.
Paring and burning, 56, 60, 73, 104, 147
Parish, as a unit for statistics, 97
Parker, R., 48 n.
Parkes, J., 143
Parkinson, John, 78
Parliamentary enclosure, 1, 50
  see also Enclosure
Pasture, see Grassland, permanent
Peacock and Handley, 40
Peakhill, 32
Pear, T., 31 n., 140 n.
Peas, 72, 73
Peat
  soils, 16, 17, 20, 21, 148, 161
  shrinkage of, 138
Pennines, 46
Permanent grassland, see Grassland, permanent
Permanent improvements, and tenant right, 135
Peter's Point, 32
Phosphorus, 162
Pinchbeck, 53
Pipes, for underdrainage, 141, 143
Plantations, 67
Plough, 151, 192
Plumb, J. H., 48 n., 190 n.
'Poaching', of clay soils, 18
Podehole, 31, 140

Poor Law, 122, 127, 192
Porter, G. R., 129
Porter, H., 40 n., 122 n.
Portland, Duke of, 74, 77
Post-glacial deposits, of fenland, 16
Potatoes, 77, 164
Prentice, J. E., 20 n.
Pressnell, L. S., 5 n., 39 n., 122 n., 193 n.
Prices, see Agricultural prices
Prince, H., 160 n.
Production costs, 41
Property Tax returns, 37, 67, 99, 194
Proto-Trent, 16
Public Record Office, 168
Pusey, Phillip, 144

Quadring, 84

Rabbit warrens, 66, 70, 104, 105, 156
Rape, 56, 72, 73, 76, 96, 108, 110, 183
Reaper, 1, 4, 150
Reclamation, 3, 23
Regional
  approach to agricultural history 8, 200,
  changes in rent per acre, 176, 180
  differences in average rent per acre, 99
  differences in crop yields, 60-1, 152-3, 185-7, 188
  differences in farm size, 91-4, 170-2
  differences in land use, 160-1, 180-3
  differences in livestock densities, 183-5
  differences in rate of improvement, 63-4, 151-4, 175-80
  differences in soil types, 17-21
  differences in underdrainage, 141-4
  differentiation, 111-13, 153-4, 175 ff., 194 ff.
  increases in rent, 36-7, 175-80
  structure of landownership, 83-8, 172-4
Regions
  agricultural, of South Lincolnshire, 95-114
  arable, of England, 47
  crop combination in the agricultural, 98
  delimitation of, 95-9
  in Fenland, 21-2
  livestock densities in, 78-81, 164-6
  of old enclosure, 50-2
  of permanent grassland, 67-9, 159
  rent per acre in the agricultural, 99

215

Index

Regions (*cont.*)
  type of farming in the agricultural, 100–10
Relief, effects of on farming, 17
Rennie, John, 27 n., 28, 29, 29 n., 31 n., 32, 32 n., 67, 140 n.
Rent, 128, 130, 134, 173, 194
  in depression, 131
  and enclosure, 39
  farmers paying from capital, 123
  as an index of growth, 35–6
  and landlord policy, 36, 63
  reduction of, 118–19, 122
  regional pattern of, 102
Rent per acre, 36, 63, 68, 108
  in agricultural regions, 95, 99, 101, 102, 103, 105, 107, 110, 112, 175, 176, 180; in England, 194–7
  factors influencing, 99
  regional pattern of, 99, 176–80
  uniformity of, 180
Rental
  accounts, of Duke of Ancaster, Sir John Thorold and Sir William Welby, 123–5
  arrears, 122, 125, 129
  increases, 35–6, 37, 107, 108, 177, 178, 179; in eighteenth-century England, 36; regional differences in, 37, 102
  receipts, 122, 124, 125
Rice, 38, 77
Riches, N., 2 n.
Richmond Commission, 166, 186
Ridge and furrow, 141
Right of entry, 133
Rippingale, 168
Rivers, gradients of and drainage, 24, 25, 29
Ross, E. D., 191 n.
Rotational grasses, *see* Temporary grasses
Rotations, 2, 3, 55, 56, 146, 182
  on clays, 147
  of open fields, 53
  see also Norfolk four-course, New Husbandry, Fallow, Improvements
Rough grazing, 103, 112, 113
  see also Waste, Land use
Roundsmen system, 127
Rural population, 192
Russell, E. J., 18 n., 19 n.
Rye, 73, 103, 162

St John's College, Cambridge, 83

St Petersburg, 46
Saunders, Joseph, 178
Scarp, 14
Schoutts, 45
Scoopwheels, 25, 29, 30
Scotland, 57, 80
Sedgebrook, 84
Seeds, 55, 70, 147, 148
  see Temporary grasses
Select Committees of House of Commons, 120, 156
Shearlings, 80
Sheep, 2, 45, 63, 71, 78, 110, 162, 194, 180
  densities, 79–80, 165, 183–4
  folding of, 18, 20, 48, 55, 58, 64, 102, 104, 105, 147, 166
  foot rot in, 64, 100, 103, 122, 142, 159, 165, 179, 183
Sheffield, 46, 80
Shorthorns, 149
Silting
  of Boston Haven, 29, 138, 139
  of Grand Sluice, 27
  of Outfalls, 32
  of Witham, 26
Skellingthorpe, 83, 119, 125, 135, 166
Skertchley, S. J., 17 n.
Skirbeck Hundred, 23, 28, 52, 53, 64, 86, 92, 99, 101, 109, 147, 148, 150, 167, 168
Sleaford, 37, 40, 43, 44, 52, 66, 91, 92, 105, 111, 122, 128, 142, 172
Sleaford Navigation, 35, 40, 44, 45
Sleaford to Tattershall Turnpike, 42, 44
Sluices, 23, 25, 27, 31, 32, 139, 140
Small farms, 64, 87, 89, 90, 91, 92, 101, 108, 151, 158–9, 166, 168, 171–5, 197
Smith, G., 20 n., 21 n.
Smith, John, 27 n.
Smithfield Market, 46, 80, 120, 149
Sod drainage, 143
Soil exhaustion, 53, 56, 59, 66, 127
Soil type, 13, 17–22, 49, 77
  and changes in cropping, 161–2, 183, 188
  and crop yields, 60–1, 111–12, 153, 185–6
  and diffusion of new farming methods, 47–8, 64, 178–180
  and land use, 66, 68, 69, 70, 71, 73, 75, 76, 147
  and old enclosure, 50

216

Index

and rent, 99
and temporary grasses, 163
underdrainage of, 141–2, 144
Somerset, 194
South Africa, 150
South America, 150
South Drove Drain, 31, 140
South-east Kesteven, 18, 67, 68, 69, 70, 83, 101, 111, 112, 122, 141, 142, 144, 153, 155, 159, 162, 171, 178, 179
  average rent per acre in, 99
  crops in, 76, 98
  enclosure in, 50
  farm size, in, 91, 93, 170, 171, 175
  farming in, 105–7
  increase of arable, 156
  lack of fertilizer, 148
  landownership in, 85, 88, 172
  land use in, 160–1, 181
  livestock density, 79–81, 164–5
  rental increases in, 37
  rotations, 147
  and transport charges, 180
South Forty Foot Drain, 25, 28, 29, 44, 139, 140
South Holland, 23, 24, 26, 30, 83, 99, 108, 109, 111, 127, 144, 148, 164, 166
  changes in transport, 180
  crops in, 98, 182
  drainage of, 31–2, 67, 140–1
  enclosure in, 50, 53
  farm size in, 170, 171
  livestock densities in, 184–5
  underdrainage in, 144
South Holland Main Drain, 25, 31, 39, 140
South Kyme, 45, 84, 139
South Witham, 15
Spalding, 25, 30, 35, 43, 44, 46, 53, 90, 108, 123, 142, 157, 164, 184
Spring
  fairs, 80
  tide, 23
Squires, absence of in Holland, 83
Stall feeding, 3, 49, 56, 145, 148, 153
Stamford, 15, 38, 40, 41, 50, 52, 108
Stamp, L. D., 13 n., 68 n., 107 n.
Steam
  engines, 137
  ploughing, 151
  pumps, in fens, 139, 140, 141
  threshing machine, 1, 4

Steeping, river, 28
Stevenson, Mr, 119
Stewart, V. F., 99 n.
Stone, Thomas, 52 n., 54 n., 55, 55 n., 56, 56 n., 57 n., 60 n., 62 n., 63 n., 67 n., 69, 70 n., 79, 79 n., 94, 94 n., 105, 152, 199
Sulphuric acid, 148
Surfleet, 172
Surrey, 134
Sussex, 134
Sutterton, 172
Sutton Bridge, 25
Sutton St Edmunds, 172
Swedes, 163, 183
Swinnerton, H., 13 n.
Sydney, Samuel, 127, 127 n.
Syston, 167

Tares, 147
Tate, W. E., 39 n.
Taxable Lands, the, 30, 53
Temple Bruer, 104
Temporary grasses, 2, 47, 48, 55, 72, 77, 146, 147, 148, 153, 163, 185
Tenancies, 62, 135
  hereditary, 135
Tenancy agreements, 69, 82
Tenant
  investment, 41, 128
  right, 132–5, 153, 189, 193
Tenants, 1, 50, 62, 64, 65, 82, 93–4, 101, 103, 129
  and agricultural prices, 135–6
  in debt, 121, 125, 172, 200
  and improvements, 85, 146
Thirsk, J., 50 n., 93 n., 100 n., 123 n., 159 n., 190 n.
Thompson, F. M. L., 36 n., 131, 131 n., 132 n., 200 n.
Thompson, Pyshey, 46 n., 56 n., 64 n., 67 n., 70 n., 71 n., 75, 77 n., 79 n., 144 n., 150, 151 n., 162 n., 167 n., 174 n.
Thompson, R. J., 36 n., 124, 124 n.
Thorold, Sir John, 83, 84, 105, 123, 124, 142
Threekingham, 36
Threshing machine, 150
Tidal waters, 22, 26, 30
Tide, high of 1811, 23
Tiles, for underdrainage, 4, 142, 143, 144

Index

Tithe
  awards, 122, 156, 173
  maps, 6, 160
Tithes, 69
Tolls, 34, 35, 41, 43
Tomes, F. M., 23 n.
Toritz, R. L., 191 n.
Torskey, 44
Torr, William, 149
Townlands, the, 31, 43, 54, 56, 71, 81, 99, 101, 112, 145, 160, 175, 180
  crop returns in 1801, 98
  drainage of, 23
  farming in, 109–10
  farm size in, 171
  increase in arable, 157, 159
  land use in, 67–9, 75, 76
  livestock in, 79–80, 164–6
  old enclosure in, 50–1
  origin of, 22
  rent in, 179
Townshend, Lord, 190
Transport
  cost in turnpikes, 43
  and declining regional differentiation, 186
Trent, river, 44, 45, 46, 68
Tritton, W., 151 n.
Trueman, A. E., 143 n.
Turnor, C., 131
Turnor, E., 35 n., 46 n., 83, 105, 172
Turnpikes, 2, 4, 18, 19, 21, 33, 34, 35, 39, 41, 42, 43, 47, 48, 49, 55, 56, 63, 64, 72, 73, 76, 77, 96, 100, 102, 103, 104, 105, 109, 111, 145, 146, 147, 163, 190
Tuxford and Sons, 151
Tydd St Mary, 68

Underdrainage, 3, 4, 18, 22, 45, 57, 76, 100, 132, 141–4, 145, 162, 178, 180, 183, 185

Vale of Trent, 13
Valuers, in tenant right, 134
Venn, J. A., 120 n.
Vernatt's Drain, 25, 30, 31, 140
Vernatt's Sluice, 30, 31
Victoria, Queen, 120

Wages, 40
Wainfleet Haven, 28
Wakefield, 46
Wales, 194, 195

Walker, J., 140 n.
War Income Tax, 102
Wash, the, 15, 17, 23, 24, 25, 32, 109, 126
Washingborough, 45
Waste, 66, 103, 156
Weaver, J. C., 96, 96 n.
Welby, 105
Welby, Sir William, 83, 123, 124, 125
Welland, river, 24, 25, 31, 32, 40, 42, 44, 45, 53, 140, 161
  outfall, 30
West Fen, 28, 39, 52, 54, 67, 70, 138, 142, 144
West Kesteven, 49, 65, 67, 70, 88, 91, 94, 99, 103, 105, 106, 107, 111, 112, 121, 141, 151, 153, 155, 171, 175, 199, 200
  arable increase in, 156, 158–9
  crops in, 75, 76
  enclosure in, 50
  farm size on, 92–3, 170, 172
  farming in, 100–2
  fertilizers, 148
  land use, 68–9, 160–5, 180–1
  livestock, 79–81, 164–5, 183–4
  occupier owners, 84–5
  relief, 15–16
  rental increases, 37, 177
  rotations, 147
  soils, 18–19, 22, 64, 178–9
  underdrainage, 142, 143
  wheat yields, 60–1, 152
West Midlands, 46
West Riding, 46, 78, 80
Wheat, 21, 49, 72, 74, 75, 96, 106, 108, 109, 110, 162, 163, 185
  prices, 34, 38, 117, 118, 119, 122, 123, 155, 158, 192
  yields, 2, 59, 60, 100, 104, 112, 117, 128, 189
Wheeler, W. H., 23 n., 27 n., 28 n., 29 n., 30 n., 31 n., 32 n., 39 n., 138 n., 139 n., 140 n., 141 n.
Whichcote, Sir Thomas, 83, 85, 106
White, C., 39 n., 40 n., 121 n.
Whittlesey, D., 96 n.
Wildmore Fen, 28, 39, 52, 54, 67, 70, 138
Willoughby, Lord, 119, 122, 130, 133, 135, 142, 143, 144, 146, 151, 170
  see also Ancaster, Duke of
Wiltshire, 190
Windmills, 25, 29, 31, 137, 139
Wisbech, 43
Witham Commissioners, 46

Index

Witham Fens, 26, 27, 67, 138, 139, 147, 148
Witham, river, 15, 16, 20, 24, 26, 27, 28, 29, 35, 42, 44, 45, 46, 52, 70, 100, 138, 140
  Act, 27, 28, 35, 39, 46
Woad, 45, 78, 164
Wolds, the, 13, 28, 56, 64, 110, 127, 152
Woodland, 103
Wool, 38–9, 46, 57, 80, 150, 178
  prices, 118, 120, 122, 159, 170
Wrangle, 52, 68
Wyberton roads, 30, 138

Yorkshire, 7, 46, 57, 78, 148
  cattle, 80
  Woad Company, 78
  Wolds, 48, 80, 193
Young, Arthur, 28, 28 n., 43, 43 n., 54, 54 n., 55 n., 56, 57 n., 62, 62 n., 64, 64 n., 66 n., 67 n., 68 n., 69 n., 76, 76 n., 77 n., 78, 78 n., 79, 79 n., 90 n., 99, 104, 104 n., 105, 152, 161, 185
Youngson, A. J., 38 n., 190 n.

www.ingramcontent.com/pod-product-compliance
Ingram Content Group UK Ltd.
Pitfield, Milton Keynes, MK11 3LW, UK
UKHW040657180125
453697UK00010B/237